Kaye M. Dan~
110 W 3d
Big Spring Texas
1 800 - 3 1 9 - 6 8 8 0

Handbook on
Petroleum Exploration

Handbook on
Petroleum Exploration

by
Suzanne Takken

with a chapter on well logging by
E.W. "Bill" Sengel

The Institute for Energy Development
Oklahoma City, Oklahoma

Handbook on Petroleum Exploration

First Printing 1978
Second Printing 1979
Third Printing 1980

Library of Congress Catalog Card Number: 78-58349

ISBN Number: 0-89419-021-0

Printed at IED Press, Inc., Oklahoma City, Oklahoma, United States of America

ABOUT THE AUTHOR

Suzanne Takken is a native of Cleveland, Ohio and a graduate of the University of Michigan with a Bachelor of Science degree in Geology. While a student she worked part-time for the U.S. Geological Survey, and from 1947 to 1970 she worked as a subsurface exploration geologist for Mobil Oil Corporation in Oklahoma City. In 1970 she became a consultant in Oklahoma City and, as such, has done consulting work in petroleum, uranium, geothermal and construction geology.

Ms. Takken is a past president of the Oklahoma City Geological Society and of the Oklahoma Section of the Association of Professional Geological Scientists. She has pursued graduate studies and continuing education courses at the University of Michigan, University of Oklahoma and University of Nevada. She has been an annual lecturer since 1972 on "Exploration Techniques for Petroleum Landman" at the University of Oklahoma, and since 1978 she has been a lecturer for The Institute for Energy Development.

TABLE OF CONTENTS

Chapter One
INTRODUCTION

Chapter Two
GEOLOGICAL TERMS DEFINED
AND ILLUSTRATED

Section One: General Terms

Section Two: Structural Traps

Section Three: Stratigraphic Traps

Chapter Three
EXPLORATION METHODS

Chapter Four
LOGGING METHODS AND APPLICATIONS

Chapter Five
MAPPING METHODS

Section One: Surface

Section Two: Subsurface

Chapter Six
CROSS SECTIONS

Chapter Seven
WHAT TO LOOK FOR IN A PROSPECT

APPENDICES

Appendix One

Appendix Two

Appendix Three

Appendix Four

Appendix Five

Appendix Six

Index

ILLUSTRATIONS

Chapter One

INTRODUCTION

A handbook is a reference manual — something to go beyond terse definitions but not to the point of being a textbook. Within that framework I have tried to include the most common terms in petroleum geology and the most common approaches to exploration. The more esoteric aspects (and there are many) will be left to some other person to describe.

Since the author is a geologist, emphasis will be on geology rather than geophysics. The latter subject is so complex that it deserves separate treatment and a complete volume of its own.

This book has evolved from a series of lectures I have given at the University of Oklahoma beginning in 1972 and continuing to the present. Neither in the lectures nor in this book is there any intention of trying to make anyone an instant expert in the field of geological exploration. Rather, the goal is to provide an understanding of the strengths and weaknesses of the exploration effort and a guide to the various methods used in that effort.

§ 1.01 Two Basic Misconceptions

Before we get into details of petroleum exploration (more precisely it should be "hydrocarbon exploration") there are two related common misconceptions which must be discussed. They have to do with the most

1

basic problems of our business and yet are so little understood by the layman that it seems appropriate to emphasize them everywhere.

First Misconception: Oil and gas are commonly thought to be in underground lakes, rivers, streams, ponds or pools. Not so. Oil and gas are in *rocks.* Or, more specifically, they are in the tiny pore spaces within rocks. We brought some of this misunderstanding on ourselves because we use the word "pool" so often. Sometimes we use the word "field" (for larger areas of production) and that is not quite so misleading. The word "pool" brings to mind some kind of large empty space filled with oil. On rare occasions large cavernous spaces do occur but they almost never have oil in them.

Oil is in the pore spaces between grains of sand, or in pinpoint pore spaces dissolved out of limestone, or in the pore spaces in the coils of a fossilized shell, or in the space opened by a fracture in the rock. But rock it is!

The porosity of a given formation ranges from zero to a maximum of about 40 percent of the rock volume. Here is how the porosity is successively reduced from original vacant space:

Imagine an empty container, such as a stainless steel tank, about 208 feet on a side and a foot high. The volume of the container is one acre-foot.

That empty container could hold 7,758 barrels of liquid. If, however, that space is filled with loose sand grains (which will later be cemented into a sandstone), the grains will occupy from 95 percent to perhaps 60 percent of the space. Assume the material takes up 85 percent of the space, which is a common occurrence in the Midcontinent, then 15 percent remains as open pore space between sand grains. Subsequent cementation of the grains of sandstone from percolating waters may reduce the percentage of pore space to 12 percent or lower. This pore space is the "reservoir" which can contain fluid.

Now, assume that oil and gas and water have migrated into the porous rock or reservoir. (All three substances commonly are found together.) Usually there will be enough water to take up a considerable amount of the pore space. A reasonable example would be a reservoir having 55 percent oil and 45 percent water.

So far we have the following figures:

7,758 barrels (original space) x
15 percent porosity (between sand grains) x
55 percent oil (in the porosity)

This calculates out to 640 barrels. This figure is the "oil in place" per acre-foot.

But there is another aspect to reservoir conditions which must be considered — the recovery factor. Much oil clings to the grain surfaces and never can be produced without special extra treatment called enhanced recovery. The primary recovery is the figure we are discussing here and that figure ranges from 10 percent to about 40 percent of the original oil in place.

Using a 20 percent recovery factor for primary production we have 640 x 20 percent or 128 barrels of oil per acre-foot. Remember that the original volume of empty space was 7,758 barrels. When we subtract the *rock*, the *water*, and the *un-recoverable primary oil*, all that remains of the 7,758 barrels is 128 barrels. (In the Appendix, the calculations of per-acre-foot recoveries for oil and for gas are restated.)

Actually, 128 barrels per acre-foot is fairly low, and the range is probably 50 to 700 barrels per acre-foot in the domestic United States. As always, the economics of any given situation are important to any drilling operation. A very low number of barrels per acre-foot is tolerable if the zone is shallow and therefore inexpensive to drill. It is also tolerable if the pay zone is quite thick and/or covers many acres. On the other hand, 300 barrels per acre-foot (BAF) can be uneconomical at great depths, or if the pay is thin, or if other expensive problems occur, such as high pressure, crooked holes, or heaving shales.

3

Second misconception: Although most people do not realize it, there is no direct way to locate oil and gas. We have no device, no method or technique which tells us, before drilling, whether oil or gas will be present. Our whole approach to the exploration for hydrocarbons is based on predicting rock formations. The assumption is that if we can identify a "trap" in rocks, then the oil and gas will be there. That is not always the case, but it is the basis for our approach to exploration.

Recently, an exception to the previous statement has been touted in geophysics as the "bright spot" technique. It does appear that bright spot analysis can identify gas trapped in some reservoirs. This can be classified as the first breakthrough in the effort to find a direct method for locating hydrocarbons. However, it has many limitations so that it cannot really qualify as a direct method. Perhaps it can be called a forerunner of direct oil-finding.

Until we have a universal tool which can "see" oil and gas directly, we shall continue to depend upon the indirect methods we have always used, which primarily deal with rock distribution patterns and attitudes and shapes.

Much of this **Handbook** will be devoted to the description of such distribution patterns and attitudes.

§ 1.02 Names of Formations or Beds

In the science of geology, there is a formal academic hierarchy of formation names and ranking. The names are derived from the nearest place-name where the formation outcrops on the land surface and has been described as the "type locality." The name often is that of a nearby creek or post office, or city. These names are officially recognized by the Stratigraphic Nomenclature Committee of the U.S. Geological Survey.

When it comes to the subsurface, a difficulty often arises as to whether the rock described on the surface as the "Wintergreen Sandstone" is the same formation as that encountered at 4,000 feet in a well a few miles away.

The oil business has handled this problem by adopting new names when there was any doubt. The common practice has been, and is continuing to be, to name oil-producing zones after the farm name on which the discovery well was located. This has been a practical way to deal with a sticky problem, although one drawback has been a proliferation of names. After years of intensive drilling we sometimes realize that the Wintergreen Sandstone and the Spearmint Sandstone are one and the same, but both names usually persist.

§ 1.03 Looking at Maps and Cross Sections

We make maps and cross sections in order to portray, to the best of our ability and our information, the three-dimensional aspects of geological formations on two-dimensional paper. A map presents the two horizontal dimensions while a cross section presents one horizontal dimension and one vertical dimension. Figure 9-B is, for example, a map view with only two dimensions. It shows a rock distribution pattern and a fault but no information is conveyed about the vertical dimension — even a half-inch below the surface. Although the map is very useful, at least one cross section, and probably several, is needed to provide a third dimension, that of depth.

A contoured map is an attempt to overcome the problem mentioned above by presenting a kind of third dimension in the form of lines of equal value. The most common such map in petroleum geology is the structure map.

One way of looking at maps and cross sections together is to imagine yourself walking along the top of some particular formation (even though it is buried several thousand feet below the surface). On Figure 1, as an example, imagine on the cross section that you are the figure walking along the top of the Green Zone. If you begin at the left side at 100 feet above sea level you will be walking uphill and when you get to the next "x" you will have gained 100 feet in elevation. You are walking along Line A-A' as shown on the map view. Each little "x" on the top of the Green Zone on the cross section is represented on the map view by another little "x" on the line A-A', which is the "route" you are walking along in the cross section.

When you get to the top of the "hill" you will be slightly higher than 300

5

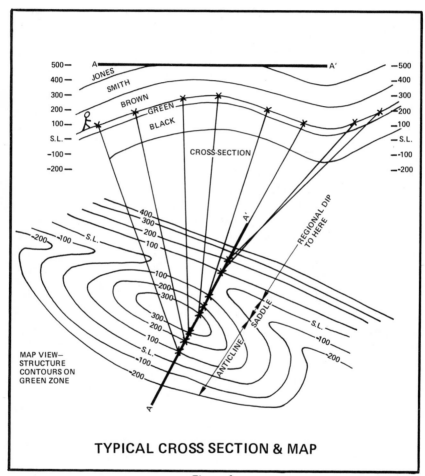

TYPICAL CROSS SECTION & MAP

Figure 1

feet above sea level and you are standing on the crest of an anticline. Then you will start down the other side of the hill. When you get down to a little below 100 feet you will find yourself going up again. The low area is often called a "saddle." As you walk upward once again you apparently are climbing another hill, but the picture ends and no other hill or anticline is defined. The straight contours on the northeast side of the feature represent the *regional dip*. An interruption in the regional dip is an *anomaly*. An anomaly can be a trap for oil and gas and, therefore, is of the utmost importance.

In the other illustrations in this book, a number of details have been omitted in the interests of simplicity but the reader should expect that most of these details will appear on prospect maps and cross sections. For instance, there should be datum plane lines (relative to sea level) on at least one side of each cross section if it is a structural cross section. These are omitted in most diagrams in the book. In addition, the background or regional contours which would normally surround the prospect anomaly have usually been omitted. Compare Figure 1 with Figure 16. They show the same anomaly but in Figure 16 the illustration is reduced to the bare essentials, because the purpose is to show shapes, not depths and thicknesses.

7

Chapter Two

GEOLOGICAL TERMS DEFINED
AND ILLUSTRATED

Section One: General Terms

The following diagrams are schematic and have no particular scale. The symbols used are those most widely used to represent certain types of rocks. A dotted pattern represents sandstone; a brickwork pattern represents limestone, and parallel closely-spaced lines indicate shale. These three rock types are the most common ones encountered in the search for oil and gas. Dolomite is another rock-type frequently encountered but it is a close relative of limestone and together they are referred to as "carbonate rocks." (Limestone is chemically calcium carbonate while dolomite is calcium magnesium carbonate.)

Each diagram is intended to show a hypothetical cross section or profile through a portion of the earth. Later diagrams also will show plan views (also called map views).

In Figure 2, three different kinds of rocks are shown in layers. We quite often call these layers beds. If the layers are fairly thick and traceable over

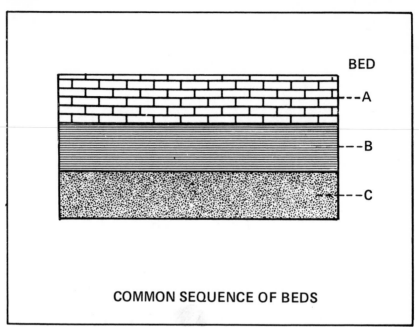

BED

--A

--B

--C

COMMON SEQUENCE OF BEDS

Figure 2

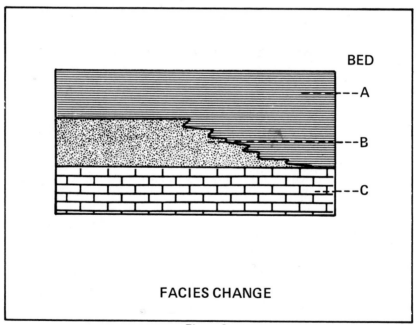

BED

--A

--B

--C

FACIES CHANGE

Figure 3

some distance, we usually give them names. There is a formal hierarchy of names in order of importance (see Introduction), but in the oil business there is a tendency to forget strict usage and call everything either a formation or a bed. Sometimes when we are not sure whether the bed being mapped is the same as the bed or layer that somebody else has named, such as the Kindblade Formation, we simply call it the A Zone or some other non-committal term. In Figure 2 the letters A, B and C are used as labels for the different beds.

Most rock materials originally are deposited on a relatively flat surface or a gently sloping one. Layer by layer, more material is added through portions of geologic time. One of the fundamental concepts of geology is that if the rocks are relatively undisturbed, the lowest beds are the oldest and each layer upward is successively younger. Therefore, in Figure 2, Bed A is younger than Bed B, which is younger than C.

§ 2.01 Facies Change

The term facies means a given aspect of rock. For instance, a single formation can have a limestone facies in one area and a dolomite facies in another; a shale may change color from maroon to green; or, in what appears to be the same layer, there can be sandstone in one place and shale in another. These lateral or vertical changes in rock type are called facies changes. They are very common, for the simple reason that no one type of material was deposited uniformly throughout the world, or even uniformly throughout the county.

Figures 3 and 4 illustrate two types of facies changes. In Figure 3 the sandstone layer (dotted) does not extend all the way across the profile, but rather is replaced by shale on the right-hand side. The zig-zag line represents the boundary of the facies change. One might imagine this sandstone to have been a deltaic deposit, of limited areal extent, with mud deposited adjacent to the sand. When each type of rock became lithified, the sand became sandstone and the mud became shale.

Another example of a facies change is shown in Figure 4. Here we have a kind of limestone mound which could be interpreted as a reef, such as a

11

coral reef. Reefs are usually rather limited in a horizontal direction but tend to grow upward and become very thick in a vertical direction. Later, other material (usually mud) is deposited on and around the reef, burying it. After lithification, it will appear that the shale adjacent to the limestone is the same age as the limestone, although it is in fact younger. In any case, the change from one kind of rock to another is a facies change. It is usually gradational, rather than abrupt, and nothing is implied as to the cause of the change. The cause is frequently unknown.

Assume a wildcat well is drilled into a limestone reef (center of Figure 4) but the fact that it is a reef isn't recognized. Oil production is established in the thick limestone zone and an offset location is staked. This second well (on the right-hand edge of Figure 4) finds only a few feet of limestone

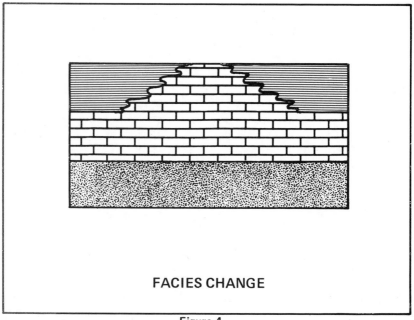

FACIES CHANGE

Figure 4

compared to the first well, and is a dry hole. The explanation for the dry hole is that it encountered a facies change and the oil reservoir was missing.

Unfortunately that explanation, while correct, is all too often used as an excuse for a dry hole and let go at that. If no effort is made to find the reason for the facies change, then the explanation is really inadequate. Careful analysis of the facies change may reveal the true nature of the trap and prevent wasteful dry holes from being drilled.

§ 2.02 Unconformity

An unconformity surface represents a gap or hiatus in the geologic record at a given locality. It can be an erosional surface or a surface of non-deposition, or both. The most widely used symbol for an unconformity is a wavy line, as shown on Figure 5. In this illustration a limestone is overlain by

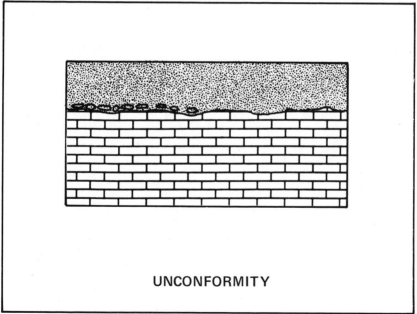

UNCONFORMITY

Figure 5

13

a sandstone which has some coarse material at its base. The wavy line indicates an unconformity and the depositional history could be reconstructed as follows:

a. the limestone was deposited in a shallow sea;

b. the sea retreated because of uplift of the land and the limestone was exposed to weathering and erosion; and

c. the land sank downward again and the sea encroached, this time carrying sand which was deposited, perhaps as a beach, on the limestone. The earliest deposits of sand included broken weathered fragments of the older limestone.

The time when the sea retreated was a time of non-deposition and also one of erosion at this particular locality, while elsewhere, during the same time, rock materials were being deposited. Thus, there can be local and even regional gaps in the geologic record. We learn to recognize these gaps simply by identifying those sections which are believed to be complete and comparing them with sections which seem to have some rocks missing. (Missing rocks also could mean faulting is present, as discussed below.)

§ 2.03 Angular Unconformity

The uplift described above is often followed by strong tilting of the rocks. Subsequent deposition occurs on the flat surface created by erosion on the tilted formations. That situation produces the angular unconformity and is illustrated by Figure 6.

Angular unconformities are usually regional in nature, often recognizable over several states. They often occur at the end of Systems (see Appendix) which are major subdivisions of the geologic column.

All unconformities have some potential as traps for oil and gas accumulation. Many large fields owe their existence to the presence of an angular unconformity; some examples are East Texas and West Edmond (Oklahoma). There is more discussion of unconformities under Types of Traps.

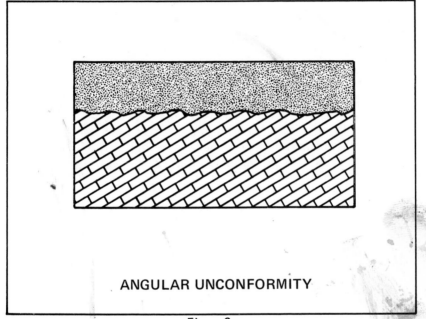

ANGULAR UNCONFORMITY

Figure 6

§ 2.04 Faults

A fault is a break in the rocks along which there has been some movement. (A break without movement is a fracture.) Sometimes the rocks on one side of the break move downward and the rocks on the other side remain stable. Sometimes the rocks on one side move up and the other side remains stable — each side can move up or down. And, there are some faults along which the blocks on either side move horizontally rather than vertically. Each of these types of faults has a name and will be described in more detail.

Since we cannot usually observe the actual movement taking place, we describe the relative movement of each fault block as up or down, away or toward the observer. In other words we deduce from the present positions of the rocks that, say, one block moved up and the other side down.

15

Normal Fault: Figure 7 is a diagram of a Normal Fault. Several layers of rocks have been broken along the diagonal line and, according to the arrows, the relative movement was down on the left and up on the right. The Normal Fault is also called a Gravity Fault because it is assumed that gravity helped to pull the left-hand block down. The amount of displacement along the fault is measured by the vertical distance between a point on one side of the fault, usually a bed boundary, and the same point on the other. In Figure 7, the base of the limestone bed on the left fault block is at A. The base of that same bed on the right-hand block is at B. The vertical distance from A to B in feet or inches is the displacement or "throw."

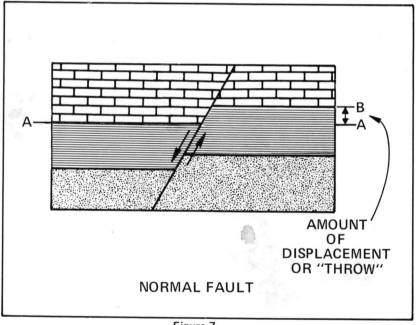

NORMAL FAULT

Figure 7

Normal faults are probably the most common type of fault in the earth's crust and the displacement ranges from fractions of an inch to thousands of feet up to perhaps 10,000 feet or thereabouts. Normal faults are frequently associated with oil and gas traps. Sometimes the fault actually is the trapping mechanism and sometimes it merely is a phenomenon to be recognized and planned for in a drilling program.

Thrust Fault: Another important kind of fault is the Thrust Fault (sometimes called Overthrust Fault or a Reverse Fault). Figure 8 illustrates a thrust fault, using the same three beds as shown in Figure 7. In this case, the result of the break in the rocks is that one block has overridden or overthrust the other. The arrows at the fault show the relative motion of each block, and the amount of displacement is the vertical distance from B to A, shown at the right.

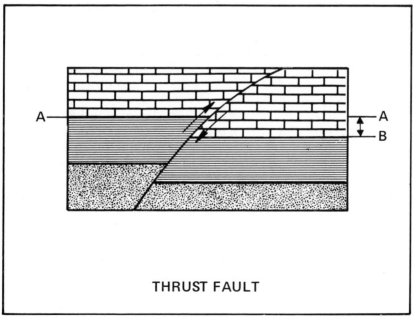

THRUST FAULT

Figure 8

Thrust faults are quite spectacular features of the earth's crust and frequently have a displacement of many miles, rather than feet. But the displacement has a strong horizontal component rather than vertical, so one block slides over the other, mostly horizontally, for great distances.

Examples of major thrust fault areas include the Appalachian Pine Mountain Thrust, the Ouachita Mountains of Oklahoma and Arkansas, and the Overthrust Belt of the Rocky Mountains.

Thrust faults, in themselves, do not trap oil and gas usually but they often overlie those traps in such a way as to conceal them from all but the most imaginative geologists or geophysicists.

Strike-Slip Faults: This is the fault along which the rocks have moved horizontally, with (theoretically) no vertical component. In reality, there often is a vertical aspect to the movement but it probably occurred at a different time. A modern term for this type of fault is Wrench Fault. These faults are not easy to identify in the subsurface, but like the Thrust Faults they can have displacements of many miles. A prominent example of a strike-slip fault which is still moving today is the San Andreas Fault of California. It extends on land from the Gulf of California to about Mendocino, in Northern California. It then goes out to sea as a part of the plate boundary system of global tectonics. It is estimated that the land on the west side of this fault has moved about 350 miles northward with respect to the east, or continental side. Of course this movement took place over long periods of geologic time, but at least one movement we know about, the San Francisco Earthquake, amounted to displacements of over 20 feet in some places.

Figure 9 illustrates a strike-slip fault in cross section (9-a) and in two map views (9-b and 9-c). In the cross section there is no visible change in the rocks from one side of the fault to the other. It is customary to use the letters A (for away from the observer) and T (for toward the observer) to indicate the relative movement along the fault plane. In the map view, 9-b, it is seen that the limestone formation has been offset by movement along the fault so that the bed boundaries X and Y on one side do not carry directly

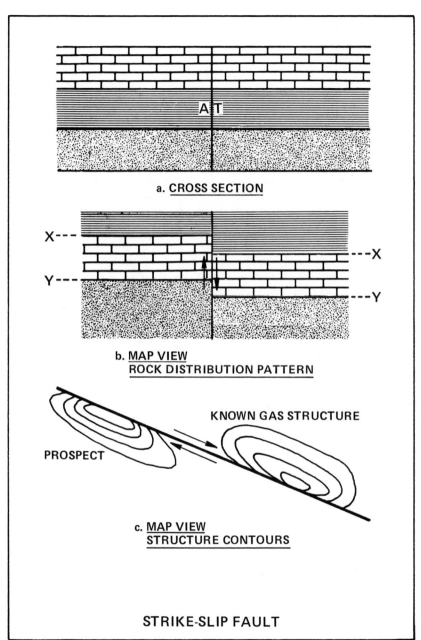

a. CROSS SECTION

b. MAP VIEW
ROCK DISTRIBUTION PATTERN

KNOWN GAS STRUCTURE

PROSPECT

c. MAP VIEW
STRUCTURE CONTOURS

STRIKE-SLIP FAULT

Figure 9

across as they should if there were no fault. Instead, the X and Y boundaries on the right side (east side) have been moved southward some distance.

The recognition of such faults is not easy, but once identified certain uses in petroleum geology are evident. Figure 9-c illustrates the idea. Assume that a strike-slip fault has been identified as a long regional fault of some 100 miles length, with a displacement of 10 miles. A known gas structure is producing at one locality along the fault and the structure contours suggest that the feature is incomplete. The place to look for the other part of the structure is 10 miles away on the other side of the fault.

One must know the direction of relative movement of the fault in order to prospect in this way. The terms for the direction of relative movement are right-lateral displacement and left-lateral displacement. This means that if one were to stand on one side of the fault (either side) the opposite block, across the fault, would have moved to the right relative to the block being stood on, in the case of right-lateral displacement. For a left-lateral fault, the movement would be to the left.

It is expected that strike-slip faults will play an increasingly important role in petroleum exploration as more detailed subsurface information accumulates.

Contemporaneous Faults: The term "contemporaneous" refers to the fact that the fault was active simultaneously or contemporaneously with the deposition of sediments in the area. One result of this is that the beds on the downthrown side of the fault are thicker than the same ones on the upthrown side.

Contemporaneous faults are a special form of normal fault. They commonly occur in the Gulf Coast area and many oil and gas traps are related to them. Figure 10 illustrates a contemporaneous fault. The fault takes an arcuate form, usually down-to-the-coast, and tends to create a "rollover" of the beds near the fault plane. This rollover can be a trap for hydrocarbons. Note that beds A, B and C are thinner on the left (high) side of the fault than they are on the right (low) side.

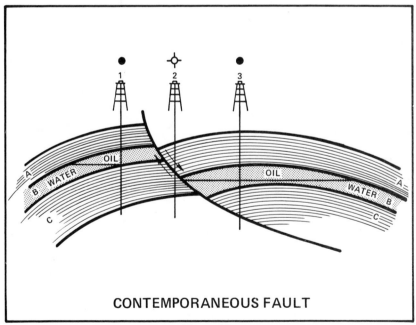

CONTEMPORANEOUS FAULT

Figure 10

In recent years, contemporaneous faults have been recognized in interior basins of the United States and they may prove to be more common than was previously supposed. Another term for contemporaneous fault is "growth fault."

Bedding-Plane Faults: This term refers to the fact that the fault seems to have no observable roots. It is usually a thrust fault where it can be observed but at greater depth there is no evidence for it and it is thought to die out along formation boundaries or bedding planes within a formation.

Figure 11 shows a bedding-plane fault. The sandstone formation has been broken and displaced in typical thrust-fault style. An anticlinal fault trap is evident on the left side of the fault. If production is established in the sandstone bed, many people would assume that a similar anticline exists at

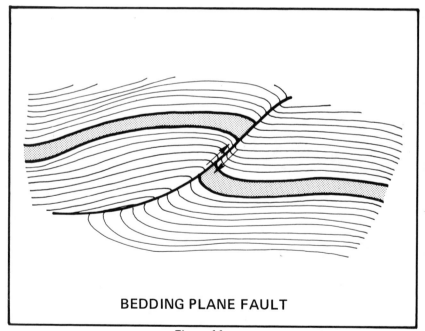

BEDDING PLANE FAULT

Figure 11

greater depth (because that is usually the case) and would drill a deep test, expecting additional production. If the fault is a bedding-plane fault (also called a decollement or detachment), there will be no anticline at depth.

This kind of faulting has been recognized in areas such as the Appalachian Mountains, where the term "thin-skinned tectonics" has been used to describe rootless faulting.

Reverse Faults: Figure 12 illustrates a reverse fault in two diagrams because a reverse fault often happens in two stages. Essentially, a reverse fault is a high-angle thrust fault. Long ago, an arbitrary dividing line between thrust and reverse faults was put at 45° of inclination. If the inclination (or dip) of the fault plane was less than 45° it was called a thrust fault. If more than 45°, and thus steeper, it was called a reverse fault. One of the problems

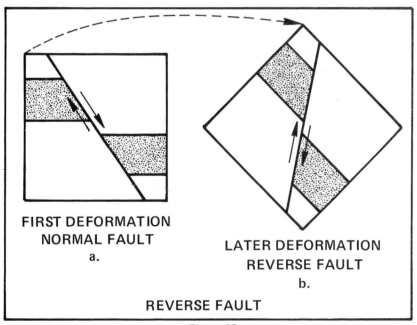

FIRST DEFORMATION
NORMAL FAULT
a.

LATER DEFORMATION
REVERSE FAULT
b.

REVERSE FAULT

Figure 12

of naming things hinges on whether the name should be purely descriptive ("that's the way it is now") or genetic ("that's how it happened"). In geology we still have, and use, a mixture of the two methods and some confusion still exists because of it.

In Figure 12, the "a" portion shows a plain normal fault which occurred as early deformation of the rocks took place. Later in geologic time, further deformation tilted the entire area, faults and all, over to the east, as shown in 12-b. These illustrations are very simplified because probably there would be other changes during the second deformation. For instance, there might have been additional movement along the fault plane, crumpling of the rocks near the fault, and so on. The idea is that the normal fault has been "reversed" by subsequent movement and now is in a high-angle thrust position, but it should be called a reverse fault.

The problem with naming or labeling faults is that we see only a portion of the fault, not its entire extent. We are forced to interpret an elephant by feeling its tusks, so to speak. Figure 13 shows what can happen along the entire vertical length of a fault. It appears to be a low-angle thrust fault in its upper portion but is a vertical normal fault at depth. Variations in the dip of a fault plane are not unusual over large distances or depths, but any small segment of the fault plane may be quite straight. Changes in dip of the fault plane occur partly because the strengths of the rocks vary and partly because the fault may have been reactivated by new stress — perhaps from a different direction.

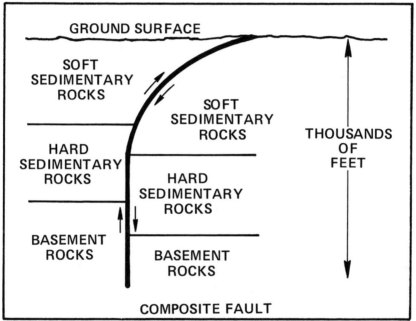

GROUND SURFACE

SOFT SEDIMENTARY ROCKS

SOFT SEDIMENTARY ROCKS

THOUSANDS OF FEET

HARD SEDIMENTARY ROCKS

HARD SEDIMENTARY ROCKS

BASEMENT ROCKS

BASEMENT ROCKS

COMPOSITE FAULT

Figure 13

Other Terms Relating to Faults: Faulted Out. This usually refers to the pay sand in some field or the main objective in a wildcat and is the explanation for the fact that the sand wasn't there. The way this happens is shown on Figure 10. Well #1 was a producer on the high side of the fault and well #3 was a producer on the low side, but well #2 was a dry hole because the fault displaced the pay sand so much that the well could not and did not encounter it. That is how a sand becomes "faulted out," and that is why field-development geology can be a tricky business in a severely-faulted structure.

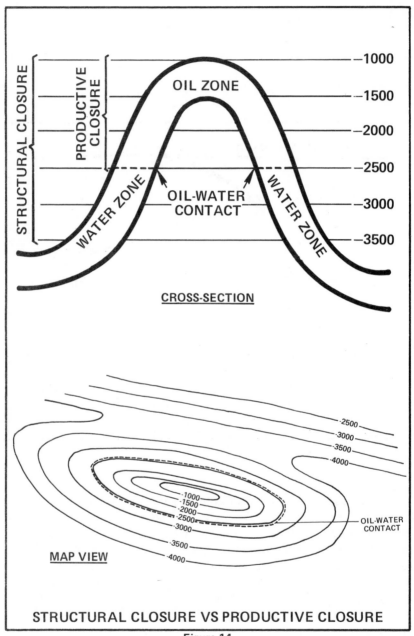

STRUCTURAL CLOSURE VS PRODUCTIVE CLOSURE

Figure 14

26

Fault Seals and **Fault Conduits.** Geologists are fond of using faults to explain both the migration of fluids and the trapping of those fluids. Although it is true that each situation indeed can occur, it is my own opinion that faults are more often seals against migration of fluids rather than conduits for fluid flow.

§ 2.05 Structural Closure vs. Productive Closure

Any trap has some measurable closure, defined either by closing contours or by some barrier such as a sand pinchout. This closure defines the area which could be filled with oil and/or gas. In many cases the structural closure also contains some water, so that the *productive* closure is less than the *structural* closure.

In the illustration the structural closure amounts to 2,500 feet (from -1,000 to -3,500), at which point the fluids would spill out and migrate farther away if any pressure differential occurs. But the productive closure is only 1,500 feet (the interval from -1,000 to -2,500 feet). All below contains water.

In map view the same situation would look like this:

The dotted line indicates the oil-water contact, which in this instance is at minus 2,500 feet (also see cross section). Thus, the productive closure is the number of feet in the structure above the contact with water. In the illustration this is the interval from minus 2,500 to minus 1,000, or 1,500 feet total. The structure is larger than 1,500 feet and could hold more oil and/or gas. The hydrocarbon supply migrating into this particular structure was apparently limited, so that water filled the remaining lower space in the structure.

In a stratigraphic trap, the situation is similar but the terms change slightly. The term structural closure is not used, since no structure exists in

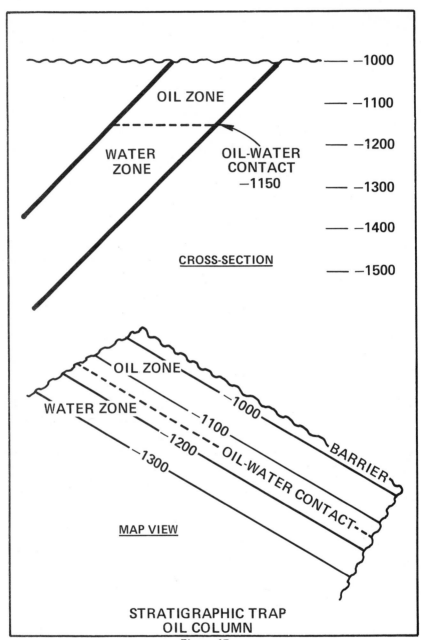

STRATIGRAPHIC TRAP
OIL COLUMN
Figure 15

the traditional sense. Productive closure now is called productive column, for the simple reason that although closure does not exist, an oil "column" does. The following sketches illustrate this point in Figures 15-a and 15-b.

Section Two: Types of Traps

Oil and gas have to be trapped, somehow. They exist, along with water, in porous reservoir rock, and they tend to move or migrate, because of differential pressures, until they cannot move any more. A trap is simply a barrier to further migration and, as such, it takes many forms. The following descriptions include the most common forms of traps. There are other complex combinations and exotic forms of traps which are omitted here for the sake of brevity. All of the illustrations which follow have a map view and a cross section for each trap.

§ 2.06 Anticlines
The anticline is one of the commonest kinds of traps. The simplest form, shown in Figure 16, is a symmetrical, usually elliptical shape about two to three times longer than it is wide. In the map view the contour lines are spaced evenly on all sides. The numbers on the contour lines usually represent feet, but the value could be meters or some other measurement.

The cross section shows, more clearly than the map, to the uninitiated, that the anticline is simply a fold or up-bending in the rock formations. The names are arbitrary ones given to various formations, just as they are assigned arbitrarily in local areas.

Figure 17 shows an asymmetrical anticline. This is quite common and is different from the first illustration in that one side (or limb) of the anticline is much more steeply dipping than the other. In the cross section the right-hand limb has a very steep dip while the left side is more gentle. On the map

29

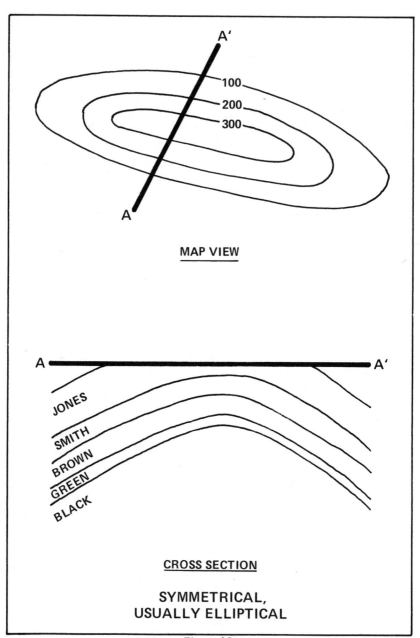

MAP VIEW

CROSS SECTION

**SYMMETRICAL,
USUALLY ELLIPTICAL**

Figure 16

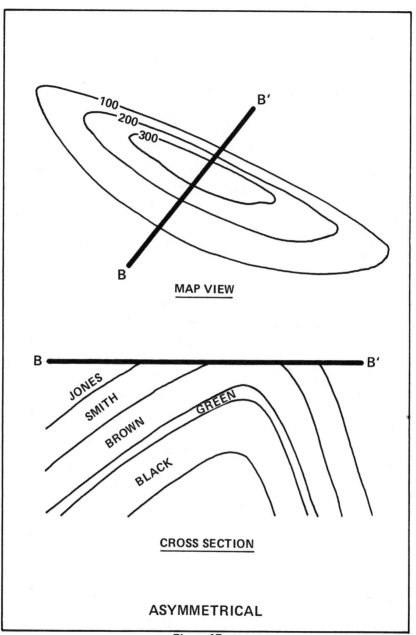

MAP VIEW

CROSS SECTION

ASYMMETRICAL

Figure 17

31

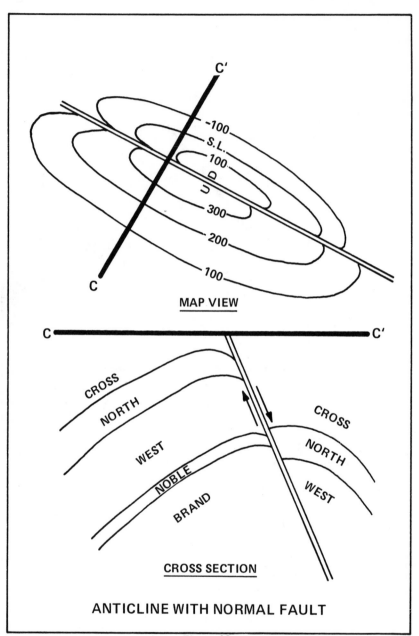

MAP VIEW

CROSS SECTION

ANTICLINE WITH NORMAL FAULT

Figure 18

view, this situation is reflected in closely-spaced contours on the steep side and wider spaced contours on the gentle side.

Figure 18 shows a faulted anticline. Without the fault it would look similar to Figure 16, but the fault has broken the structure and allowed one side to slide down with respect to the other. "U" and "D" mean "Up" and "Down" and describe the relative movements on either side of the fault. Here the contours on the map view again represent feet, both above and below Sea Level (which is our only common denominator). On the downthrown side of the fault, the term S.L. (for Sea level) has been used instead of arithmetic zero which would also be appropriate. Either form is correct, but S.L. is more precise.

Figures 19, 20 and 21 show more complex anticlines because they are effected by faults or salt intrusions or both. An overthrust anticline is shown in Figure 19. In the cross section view, it will be seen that Zones 1 through 4 have been pushed up over their counterpart zones. A well drilled near the center of the cross section would encounter Zone 3 twice and Zone 4 twice. On the map view, an anticline is shown on one side of the fault and just plain dip is shown on the other. Oil and gas can be trapped on either side, however. On the high or overthrust side, the trap is anticlinal. On the downthrown side, the fault forms the trap. So this case is a combination of anticline and fault trap. Note that the contours have minus signs, indicating they are below Sea Level, and therefore the lowest number (-6,000) is the highest part of the structure.

Figure 20 represents a piercement salt dome. Any dome is really a more circular version of an anticline. The cause of this dome is the upward movement of a large salt plug, and it is called piercement because it actually penetrates up through many formations. The upward movement also causes rather complex fault patterns in the rocks overhead. Often the faults are branching, and to determine the strike and dip of each branch can be a development geologist's nightmare.

Sometimes salt intrusions do not reach close enough to the surface to be penetrated by the drill. They cause an arching overhead, nevertheless, often accompanied by a central graben or down-dropped fault block.

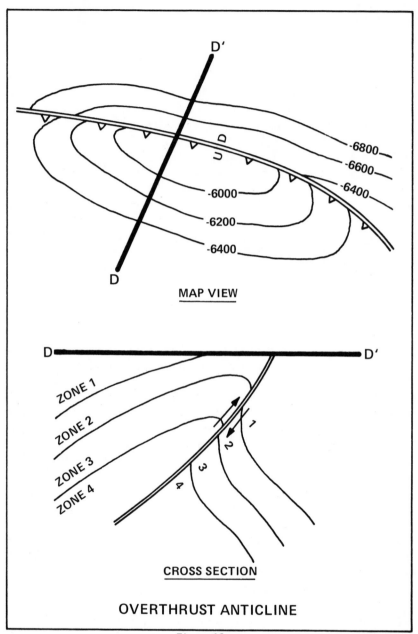

D'

D

-6800
-6600
-6400
-6000
-6200
-6400

MAP VIEW

D ─────────────────────── D'

ZONE 1
ZONE 2
ZONE 3
ZONE 4

1
2
3
4

CROSS SECTION

OVERTHRUST ANTICLINE

Figure 19

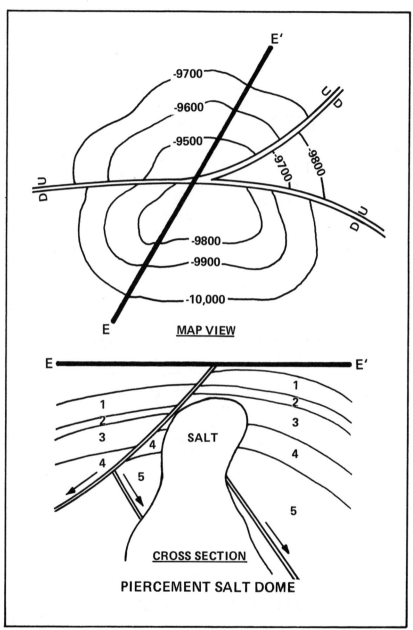

E'

-9700

-9600

-9500

D U

D

-9700

-9800

U D

-9800

-9900

-10,000

E

MAP VIEW

E ●————————————————————————● E'

1

2

3

1

SALT

4

2

3

4

5

4

5

CROSS SECTION

PIERCEMENT SALT DOME

Figure 20

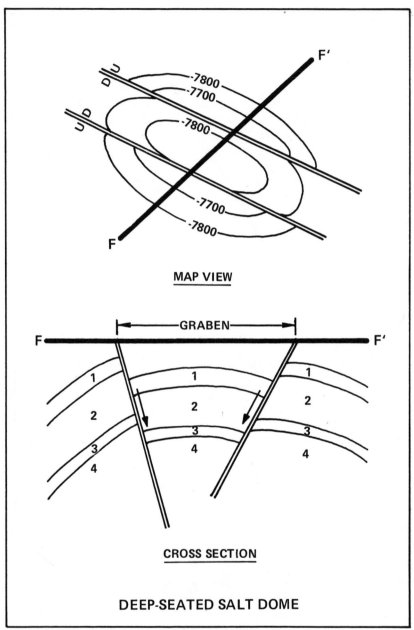

MAP VIEW

CROSS SECTION

DEEP-SEATED SALT DOME

Figure 21

Figure 21 illustrates this type of feature. On the cross section the graben is defined by two normal faults, dipping toward each other. This graben is really a shallow wedge and does not extend to great depths, but all zones in the graben are structurally lower than the same zones on opposite sides of the fault.

On the map view, the appearance is of an anticline crossed by two faults. Its roughly circular appearance and location in salt-dome territory suggest that the cause of the feature is a deep-seated salt dome. Salt domes occur in a wide belt along the Gulf Coast and are responsible for many sizeable oil and gas accumulations.

§ 2.07 Fractures

Fractures, also called joints, are present almost universally to some degree in any area which has undergone tectonic movement. Sometimes they occur in association with faults, sometimes not. They may be more common in carbonate rocks than in shales or sandstones, but can be present in all types. Some rocks, particularly some shales, tend to be rather plastic and are more apt to flow, under pressure, than break. But most rocks are quite brittle and undergo fracturing quite easily.

Artificial fracturing, through the well bore, has been commonly used for many years to enhance production from tight pay zones. More recently, massive hydraulic fracturing has been introduced to stimulate production from reservoirs (mostly carbonates) which almost certainly would have been abandoned a few years ago. The cost of these massive frac jobs could not be justified by $3.00 oil, but at today's higher prices they are feasible.

It is not certain what role natural fractures play as reservoirs. Most fractures seen in cores are either tightly closed or filled with a secondary deposit such as calcite. It is not common to see open fractures which might represent reservoir space. Nevertheless it is almost certain that such space does occur. Although most fractures occur on a relatively small scale — for instance one might see 30 or 40 fractures in a core you can hold in your hand — and they can occur on a grand scale too. The Albion-Scipio Field in Michigan, for example, is considered to be a large-scale fracture zone. It is approximately 35 miles long and the producing formation is a secondary dolomite which has replaced the original limestone along the fracture trace.

Fractures fall under the category of structural features. Their role as reservoirs is obscure, but their presence is almost universal.

§ 2.08 Faults and Fault Blocks

Faults have already been described at some length under "General Terms Defined and Illustrated," including something about the trapping aspects. However, a further discussion in this section is to emphasize the fact that faults are structural features and that they are common occurrences on most anticlines or domes. They may help to trap oil in certain portions of the structure and may allow it to leak out elsewhere. Faults may not have any significant trapping effect at all, but they serve to complicate the geology so that prediction is difficult.

Fault blocks are portions of a general area which are broken by faults, so that some blocks are uplifted and some are downwarped. An uplifted block is often called a *horst,* and a downthrown block is often called a *graben.* These are German words which are used widely in geology. These individual blocks may serve as traps for oil and gas (usually in the upthrown or horst block) because they are structurally high to the region even though they may not be anticlinal in form.

On salt domes, as an example, some fault blocks are more nearly like pie wedges. They are simply separate segments of the overall "pie" but may have different producing characteristics. So the term "fault block" refers to a portion of land which can be defined by faults on at least three sides and must be dealt with as a separate entity from other fault blocks. The geologist and the reservoir engineer must recognize the possible fault blocks in order to deal with them effectively.

Section Three: Stratigraphic Traps

Most traps that we call stratigraphic are really combinations, in that they have to have some structural situation in conjunction with a stratigraphic anomaly in order to form a trap. In other words, there must be regional or local tilting of some kind before a pinchout or unconformity zone can become a trap. On the other hand, there are "pure" stratigraphic traps also. These are in no way dependent on structural position and can occur in perfectly flat terrain. Examples would be reefs, channel sandstones and deltas.

Stratigraphic traps have a great deal of variety, compared to structural traps. They are also more subtle, and therefore more difficult to find. And they tend to offer only one formation yielding oil and gas versus several formations, commonly, in structures. One other aspect is important to note: they can cover an extremely large area.

In the United States, except for Alaska and certain offshore areas, I think it is fair to say we have already discovered perhaps 85 percent of the major structures which will produce oil and gas. But my estimate for stratigraphic traps is that we have discovered only about 40 percent so far. These estimates are purely intuitive, and are not based on any statistics or any formal estimate of reserves.

Thus the future for new hydrocarbon discoveries in the United States depends tremendously on our ability to locate stratigraphic traps. Most of them in the past have been found accidentally (usually while drilling in search of a structure), but now we must make a deliberate effort to seek out and drill the ever-nebulous stratigraphic traps on their own merits.

§ 2.09 Reefs

One of the most outstanding examples of a pure stratigraphic trap is the reef. Coral reefs, as they are often called, even though corals may not be a major constituent of the reef, are found world-wide today, in places such

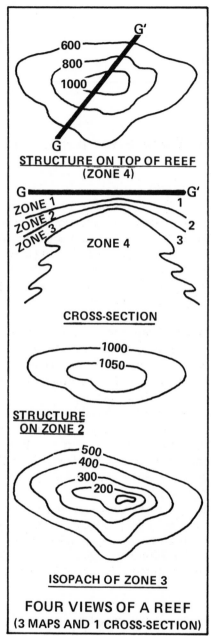

STRUCTURE ON TOP OF REEF
(ZONE 4)

CROSS-SECTION

STRUCTURE
ON ZONE 2

ISOPACH OF ZONE 3

FOUR VIEWS OF A REEF
(3 MAPS AND 1 CROSS-SECTION)

Figure 22

40

as the Bahamas, the Florida Keys, and the world-renowned Great Barrier Reef of Australia. Geologically, old reefs are found throughout the stratigraphic column and are important reservoirs for oil and gas.

In the subsurface, a reef often takes the form of a mound — a pile of limestone made up of fossil fragments of all kinds. It is usually surrounded by shale. The vertical dimension is quite large compared to the horizontal, and this is why they are sometimes called pinnacle reefs.

Figure 22 shows a diagrammatic cross section of a reef and three different map views. In the profile, the reef is Zone 4. Note the mound-like shape and the irregular sides, and that the whole thing is surrounded by Zone 3. The deposition of Zone 3 lasted only long enough for a little of it to be deposited on top of the mound. It also has filled in all around the mound. By the time Zone 2 was being deposited, differential compaction of the wet clays of Zone 3 allowed the material to be depressed on the sides of the reef. Therefore, Zone 2 and to some extent Zone 1 which follows, show an arching effect over the mound. A structure map of the reef surface shows relief of about 500 feet, which would be the height of the reef above the sea-floor when it was a living reef (unless later erosion has worn it down). Compare that structure map with one on the top of Zone 2. The relief, or closure, amounts to only about 75 feet.

Assume for a moment that no wells had been drilled deep enough to encounter the reef, but that several had penetrated Zones 1 and 2. Careful mapping would reveal what appears to be an anticline. Deeper drilling would reveal the reef. Also, the rather dramatic thickening of Zone 3 could be a clue. Suppose a wildcat well was drilled which, although on the crest of the reef, did not have the reef as an objective, but rather Zone 2. The well penetrated perhaps a few feet into the reef and was plugged. If any other wells nearby had a very thick Zone 3, a thoughtful geologist might conclude that there is a potential anomaly in the area.

Reefs can take many shapes. Atolls, for instance — the circular shape with the open center. A good example is the Horseshoe Atoll of West Texas (Figure 23), which includes the Scurry County Reef (Figure 24) and several

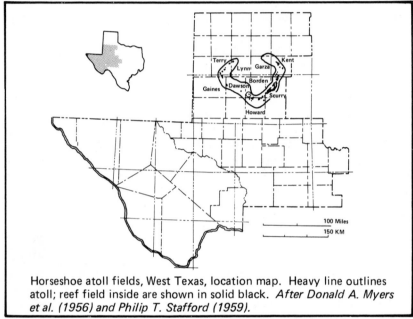

Horseshoe atoll fields, West Texas, location map. Heavy line outlines atoll; reef field inside are shown in solid black. *After Donald A. Myers et al. (1956) and Philip T. Stafford (1959).*

Figure 23

others. Some features in the subsurface actually are banks of reef debris rather than true reefs. They take an elongate form, much like a channel sandstone. The two principal differences are in rock type and in orientation. Lime banks generally will follow regional strike whereas channel sands will, as streams do, move down slope.

Because of their varying shapes and abrupt boundaries reefs are difficult to explore for. Seismic work, in recent years, has been successful in locating pinnacle reefs, particularly in Michigan. But, seismic reflections cannot always "see" lime banks and atolls partly because of the relatively subdued nature of the reefs in relation to thousands of feet of adjacent rocks which may make themselves more visible as seismic "events."

The pore space in a reef is extremely variable. Some animals and plants

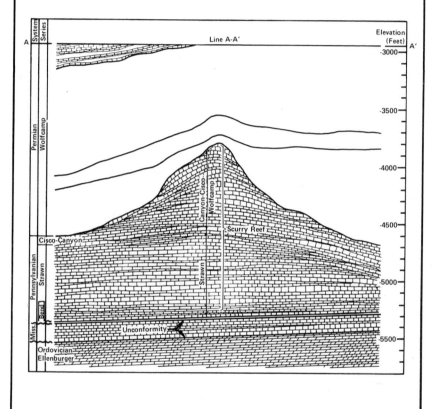

Kelly-Snyder oil field, cross section along line A-A'. West-east section through Scurry reef. *After Philip T. Stafford (1957).*

Figure 24

43

have very open skeletal remains, like latticework. Others, while having very solid shells like mollusks, may fall in the debris in such a way as to leave a large opening under the shell. If sediment and limy ooze do not fill up too much of this pore space, it will be available for hydrocarbons. Later changes (diagenesis) can alter the volume of porosity also. The chemical replacement of limestone with dolomite enhances the porosity somewhat. However, a related material, anhydrite, can plug up a lot of pore space. If the reef also has been subjected to fracturing, this could enhance the porosity and also the permeability.

Because of the difficulty in determining accurately the porosity and permeability of a reef, reserves of hydrocarbons also are very difficult to calculate in a discovery well. Even with several wells drilled there can be a wide range of estimates. Actually, these difficulties extend to almost any limestone reservoir, particularly if it is fractured. About the only exception is oolitic limestone which usually has its porosity and permeability rather evenly distributed and is therefore more like a sandstone.

Another difficulty in calculating reserves in any stratigraphic trap is that the areal extent of the producing zone is not known early in the game. One really needs some production history so that decline curves can be constructed before reasonably accurate reserve estimates can be made. Another way is to find a fairly similar older field and assume that the reserves of the new field will be like the older one. There is more discussion of reserves in Chapter Six.

§ 2.10 Irregular Sand Bodies

Grouped under this category are such features as stream channels, offshore bars, deltas, beaches, dunes, and similar forms which are quite limited in at least one direction. For simplicity, anything less than a sheet sandstone would fall into this grouping. These bodies are, for the most part, pure stratigraphic traps also (as is the reef) but may have some component of regional or local dip affecting the hydrocarbon accumulation.

Stream Channel. In the cross section of Figure 25, a channel sand, labeled the Smith Sand, is shown. It is a limited sandstone body completely

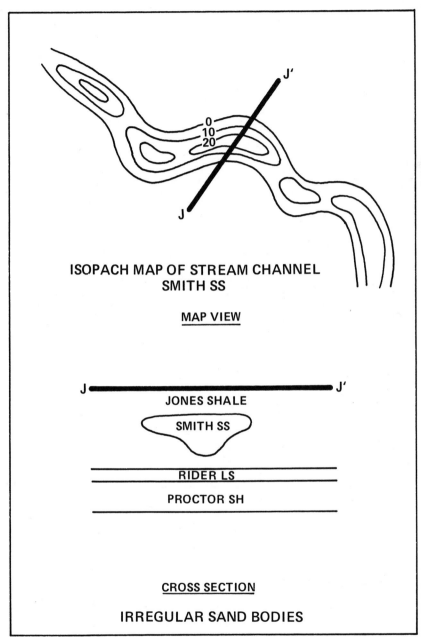

ISOPACH MAP OF STREAM CHANNEL
SMITH SS

MAP VIEW

CROSS SECTION

IRREGULAR SAND BODIES

Figure 25

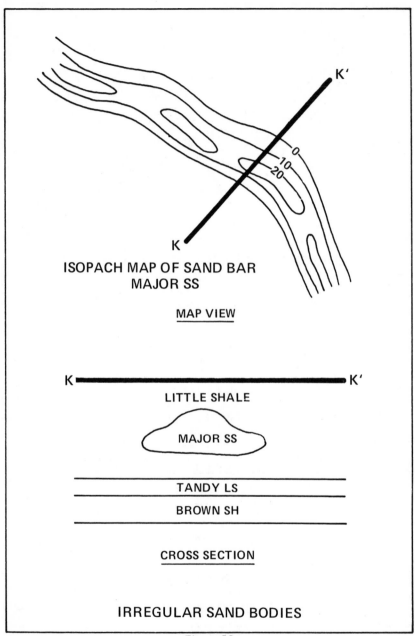

ISOPACH MAP OF SAND BAR
MAJOR SS

MAP VIEW

LITTLE SHALE

MAJOR SS

TANDY LS

BROWN SH

CROSS SECTION

IRREGULAR SAND BODIES

Figure 26

enclosed in the Jones Shale. The formations above and below are flat-lying, although there is probably some dip of all the formations somewhere in the area. The profile of the channel is characterized by a fairly flat upper surface and a lower surface which is convex downward. The scouring action of the stream cut an early, narrow channel and a later broader channel.

To make a structure map on top of the sand body, or of any nearby formation, would not be informative. The most appropriate map is an isopach (thickness) map of the sand body itself.

In Figure 25 the isopach map indicates a meandering shape such as streams tend to take, and the boundary of the channel is defined by the zero contour. According to the contour numbers, the maximum thickness is approximately 25 feet (just a bit more than the highest-value contour of 20 feet). The width or the length is not identifiable because no scale is shown, but the proportion of the width to the length is roughly appropriate. It is common to have pods of slightly thicker sand rather than to have a consistently thicker center portion.

Figure 26 shows a cross section and isopach map which at first glance seem very similar to Figure 25. They are indeed similar but not identical. In the profile the flat-lying formations are the same with different names. The sand body, the Major Sandstone, has a reversed shape from the one in Figure 25. The base is relatively flat while the upper surface is convex upward. This is characteristic of an off-shore bar. Neither of these two types of features is as symmetrical in actuality as they have been drawn here, but the overall differences in shape are correct.

The isopach map of the Major Sandstone is almost identical to the isopach of the Smith Sand except for fewer meanders.

In essence, this means that an isopach map does not clearly show the geometry or vertical shape of the sand body being mapped. Again, the information is only two-dimensional. One needs both a map and a cross section to get the three-dimensional picture.

Galveston Island is an offshore bar and so is Padre Island, both in the Gulf of Mexico. Galveston Island particularly has been extensively studied by geologists to learn the details of sediment deposition of such bars.

Deltas. Most geologists now believe that the deltaic environment, with its many varied aspects, is the source area as well as the reservoir area for an enormous amount of hydrocarbon origin and accumulation. In other words, the oil was born there and didn't travel very far to find a permanent home. Therefore modern deltas are also being intensively studied in order to understand geologically old deltas.

Deltas take many diverse shapes. Two of the most famous are the Nile Delta with its classic triangular shape for which the name delta was an obviously appropriate term, and the Mississippi Delta, again with an appropriate name — the Birdsfoot Delta.

Figure 27 shows, in cross section, the very irregular edges of a deltaic mass and the relatively thick center portion. The map view also shows the irregular nature of the feature. Delta deposits consist primarily of clastic materials, that is sand and clay. The clay, being much finer, is carried out farther to sea while the heavier, coarser sand is deposited closer to shore. Channels also develop in the deltaic environment. They are the major distributaries of the river. They form the passes which ships take from New Orleans to get to the Gulf of Mexico.

Other aspects of deltas include the inter-deltaic environment, the effects of longshore currents in redistributing the sands, the mud flats, lagoons, mud-lumps, and so on, all of which make up the total system and which make it very complex. Thus, it is easy to see why subsurface deltas are quite difficult to recognize and interpret.

A classic example of the interpretation of a subsurface delta is the study by Busch (1959) on the Booch Sand of eastern Oklahoma. Figure 28 is his map of the Booch Delta. Another example of the interpretation of an entire system was done more recently by Visher & Saitta on the Bluejacket Sandstone, again in eastern Oklahoma, shown in Figure 29. Busch's map is a standard contour map with a deltaic interpretation. The Saitta map is of special interest because of the use of electric log curve shapes to aid in the interpretation. The electrical responses of different types of sand bodies can be used in this fashion, if there is sufficient well control in the area under study.

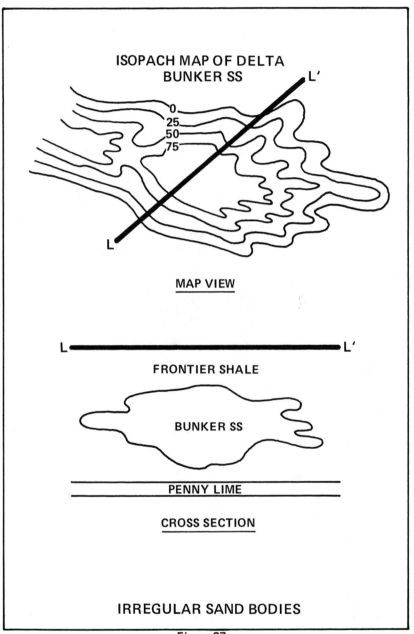

ISOPACH MAP OF DELTA
BUNKER SS

L'

0
25
50
75

L

MAP VIEW

L ————————————— L'

FRONTIER SHALE

BUNKER SS

PENNY LIME

CROSS SECTION

IRREGULAR SAND BODIES

Figure 27

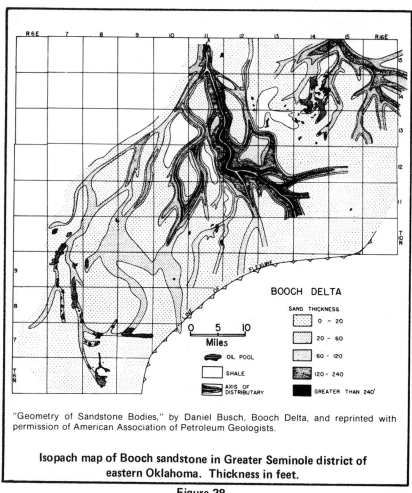

"Geometry of Sandstone Bodies," by Daniel Busch, Booch Delta, and reprinted with permission of American Association of Petroleum Geologists.

Isopach map of Booch sandstone in Greater Seminole district of eastern Oklahoma. Thickness in feet.

Figure 28

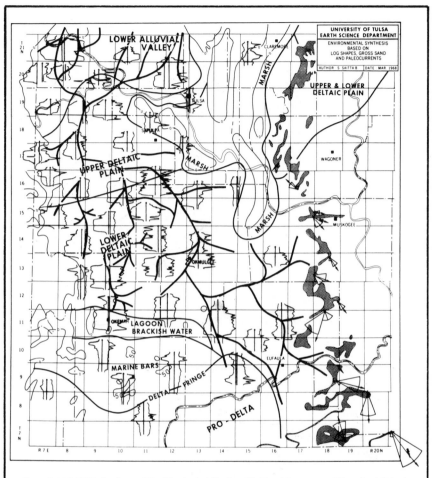

Copyright © 1968, Geology of the Bluejacket-Bartlesville Sandstone, page 59, and published with permission of Oklahoma City Geological Society, Inc.

DEPOSITIONAL ENVIRONMENTS

Following the deposition of the Brown Limestone and a thick shale overlying it, a major regression took place in the area of eastern Kansas and Oklahoma. During this time streams carrying terrigenous materials cut into the subaerial portion of the uplifted continental shelf and formed a major fluvial channels. As these channels prograded seaward the large Bluejacket deltaic complex was deposited. This deltaic progradation produced six environmental units: (1) lower alluvial valley; (2) upper and lower deltaic plains; (3) delta fringe and pro-delta areas; (4) marine bars; (5) lagoons and bays; and (6) marshes. Each of these units will be described and their position within the deltaic complex shown.

Figure 29

51

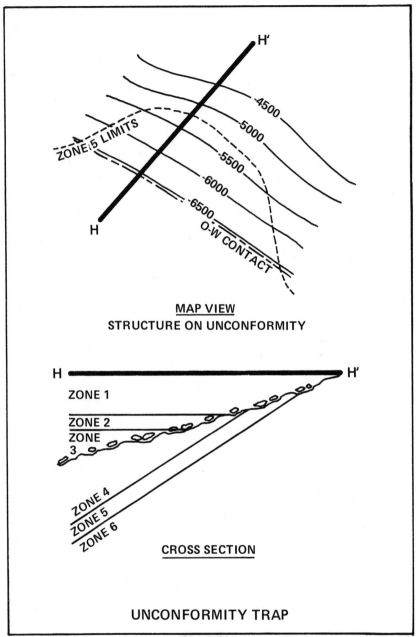

MAP VIEW
STRUCTURE ON UNCONFORMITY

CROSS SECTION

UNCONFORMITY TRAP

Figure 30

§ 2.11 Unconformities

Unconformity Trap: This is one of those stratigraphic traps which requires some kind of structural relationship to be an effective trap. It is actually a combination trap, but usually is referred to as a stratigraphic trap. This is the type of trap which can be a giant oil field. The others are usually much more limited in areal extent but this one can cover many miles. Examples are the East Texas Field and the Watonga Trend of central Oklahoma.

Figure 30, in the cross section, shows six numbered zones interrupted by a wavy line which represents the unconformity. Unconformities were previously defined in Chapter Two. In this example, Zones 4, 5 and 6 were deposited parallel to each other and probably on a relatively flat surface. At the end of the deposition of Zone 4 the region was uplifted, tilted and eroded. Then it was tilted further so that the erosional surface, which once was relatively flat, is now tilted. Following this, Zones 3, 2 and 1 were deposited in a horizontal position.

Zone 5 is a porous rock into which oil has migrated, but the unconformity surface forms a seal so that the oil cannot go upward past the seal. That is the unconformity trap.

In this illustration, Zone 2 could be another reservoir rock with the trap formed by the same unconformity. It is also possible that oil had once migrated up Zone 4 as far as the unconformity and, in this case, the unconformity did not form a seal so that Zone 2 could "thief" the oil out of Zone 4. As with faults, we sometimes use unconformities to explain traps and sometimes to explain dry holes.

Also, unconformities sometimes have a rubbly layer in their own right. Other times, the actual contact is extremely clean and Zone 1, for instance, will be resting directly on Zone 5. If Zone 1 is an impermeable rock, then any oil below will be trapped at the unconformity contact.

So far we have just examined two dimensions of this feature. Again taking Zone 5 as the pay zone, what is its areal extent? It may be many miles if it is a uniformly deposited sand, or it may be a rather localized feature. In any case it is desirable to try to map the productive limits of the zone. A

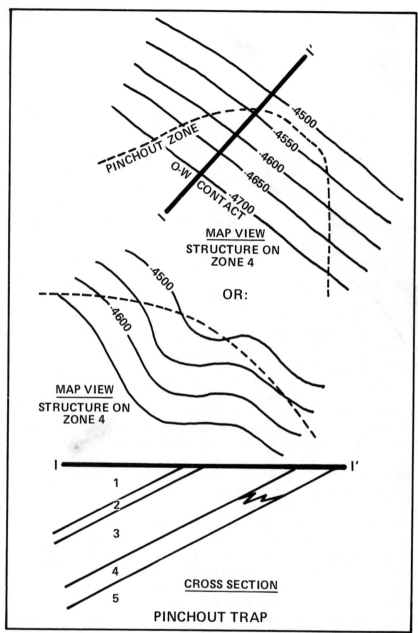

MAP VIEW
STRUCTURE ON ZONE 4

OR:

MAP VIEW
STRUCTURE ON ZONE 4

CROSS SECTION

PINCHOUT TRAP

Figure 31

structure map is not, as usual, very informative. The map view shown has contours on the unconformity itself. It would be appropriate to draw another map on Zone 5. At any rate, the updip limit of Zone 5 is where it meets the unconformity. The downdip limit of production will probably be at an oil-water contact. The lateral limits may take the form of a permeability barrier shown by the dotted line on the map. All of these limits will have to be defined by drilling.

Pinchout Trap: This is similar in many respects to an unconformity trap. There are uniformly dipping beds, that is, no anticlinal closure, and the trap is formed by an updip change in the reservoir rock itself. In Figure 31, the cross section shows five zones parallel to one another, all dipping in the same direction. Zone 4 is the reservoir rock and the zig-zag line at the right end indicates the pinchout. The term "pinchout" is a general one for several different conditions; for instance, a change from sandstone to shale. Such a change may be gradational or abrupt, depending on the type of sand body. Another change would be one of cementation whereby the updip portion of the reservoir is tightly cemented so that porosity and permeability are reduced virtually to zero.

Another example would be that of a porous dolomite lens in an impermeable limestone. When the facies change occurs — the change from dolomite to limestone — the trap occurs.

As in the unconformity trap, we need to know the lateral boundaries as well as the updip boundaries. A structure map drawn on top of Zone 4 is shown in Figure 31. No evidence of structural closure is present. The lateral boundaries will most likely be formed by lateral pinchouts similar to the updip pinchout.

Occasionally, a slight nosing will aid in the trapping. This is shown in the alternative structure map in Figure 31. A "nose" is a weak fold, not strong enough to have any closing contours. Such noses are often the chosen place to drill when all other facts are equal.

Stratigraphic traps offer problems to the explorationist which are rarely associated with structural traps. Assume a wildcat well failed to find production in its original objective, but accidentally found production in a "stray" sand. (A "stray" is an unexpected sand development usually not matching any known sands in nearby wells.) The geologist recognizes that it could be a channel sand. Or it could be an offshore bar. Possibly a beach? Each of these features would have a different areal extent and orientation. The problems are where to take additional leases, where to drill next, how to calculate reserves, and so on. One nine-inch hole in the ground is not much to go on.

Another problem is that of recognition of the type of trap. Assume a wildcat well has been drilled on a small structure, according to seismic evidence. Acreage has been leased on top of the closure and immediately adjacent as a precaution. Production is established and two offset wells are drilled. These additional wells, though productive, demonstrate that the closure doesn't exist — there is only regional dip. It is now time to reassess the geological interpretation. If it turns out that an unconformity trap is present, the productive area could be many times the original leased area, and a new leasing program should be begun at once.

There have been efforts over the past 20 years or more by prominent explorationists, such as Michel Halbouty of Houston, to direct our attention more closely to stratigraphic traps. To a large extent these efforts have failed, at least at the management level. There is still, after all these years, a tendency to look for the closure — the bulls-eye — and keep away from stratigraphic traps.

Timing is such an important factor in an exploration program. In dealing with stratigraphic traps, early recognition of the type of trap and/or type of sand body is important to the leasing program and to the development well program, so that we can proceed as economically and as efficiently as possible.

To conclude this chapter, some maps and cross sections of actual fields are presented. They are all taken from *Petroleum Geology of the United States* by Kenneth K. Landes.

Figure 32 shows the large, asymmetrical anticline of the Rangely Field in northwestern Colorado. This map is a very typical presentation in several ways. The contour lines are solid lines while the oil-water and gas-oil

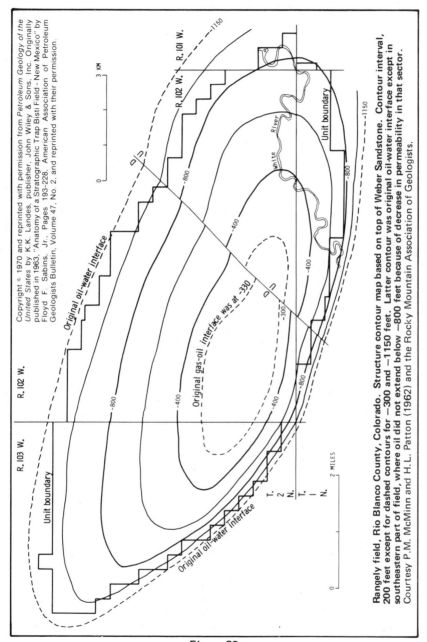

Rangely field, Rio Blanco County, Colorado. Structure contour map based on top of Weber Sandstone. Contour interval, 200 feet except for dashed contours for −300 and −1150 feet. Latter contour was original oil-water interface except in southeastern part of field, where oil did not extend below −800 feet because of decrease in permeability in that sector. Courtesy P.M. McMinn and H.L. Patton (1962) and the Rocky Mountain Association of Geologists.

Figure 32

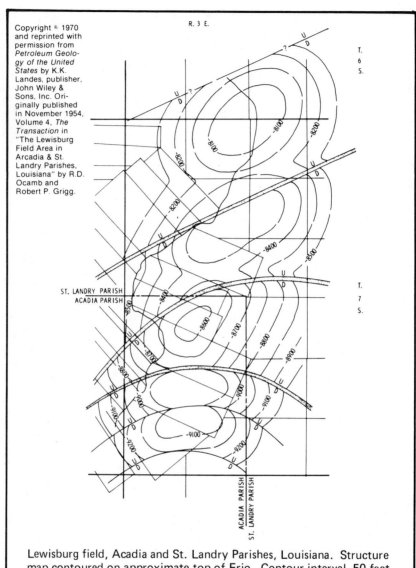

R. 3 E.

T. 6 S.

T. 7 S.

ST. LANDRY PARISH
ACADIA PARISH

ACADIA PARISH
ST. LANDRY PARISH

Lewisburg field, Acadia and St. Landry Parishes, Louisiana. Structure map contoured on approximate top of Frio. Contour interval, 50 feet. West side of T.7 S, R. 3 E. is 6 miles (9.66 km).

Higher parts of all five fault block anticlines shown are productive of gas condensate. *Courtesy Ray Ocamb and Robert P. Grigg, Jr. (1954) and the Gulf Coast Association of Geological Societies.*

Figure 33

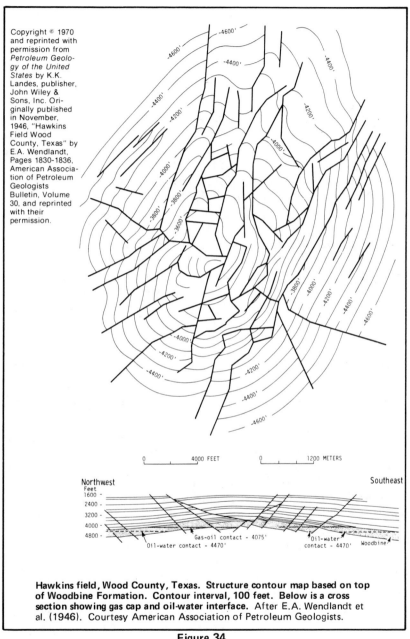

0   4000 FEET   0   1200 METERS

Northwest   Southeast
Feet
1600 –
2400 –
3200 –
4000 –
4800 –
                    Gas-oil contact - 4075'   Oil-water
         Oil-water contact - 4470'   contact - 4470'   Woodbine

Hawkins field, Wood County, Texas. Structure contour map based on top of Woodbine Formation. Contour interval, 100 feet. Below is a cross section showing gas cap and oil-water interface. After E.A. Wendlandt et al. (1946). Courtesy American Association of Petroleum Geologists.

Figure 34

59

contacts are dashed. Not all the contours are numbered, but it is understood that the lines without numbers have a value exactly half-way between the values of the adjacent contour lines. All maps are understood to be presented with north at the top, unless otherwise stated.

Figure 33 shows another anticline, in Louisiana, which is quite symmetrical with an elongate shape oriented north-south. Five faults cross the anticline, thus separating it into five smaller anticlines, each of which is productive.

The Hawkins Field in Texas is shown in Figure 34. Because of its roughly circular nature it is called a dome and it probably is located over a deep-seated salt dome. The structure is riddled with faults which do not effect the trap very much but which obviously could cause some problems in predicting the geology of development locations. In the cross section at the bottom of the illustration it will be seen that the very center block is dropped down with respect to the blocks on either side. This is the central graben so common to salt-dome features.

Figure 35 is a structure map of the area in Montana which includes the Bell Creek Field. It is obvious that there is no closure, not even a nosing, to indicate that a trap is present. This is not a map to use in prospecting but it can be used, after the field is discovered, for predicting oil-water contacts and total depths for development wells. Figure 36 more clearly tells the story of the field, which is producing from the Muddy Sand in a stratigraphic trap. Figure 36 is an isopach map depicting the net thickness of the producing zone, and the zero line defines the limits of the sand.

Figure 37 is a map and a cross section of a salt-dome field. The overhang of the upper part of the salt is clearly shown in the cross section. Some wells have drilled through the overhang. Most of the production seems to be located on the northwest side of the dome.

Figure 38 shows how complex a hydrocarbon accumulation can be. This is a gas field, South Capano Bay, and the various fault blocks, while anticlinal, are not all productive.

The Carter-Knox Field of Oklahoma, shown in Figure 39, is an anticline in which the deep portion is much displaced from the shallow

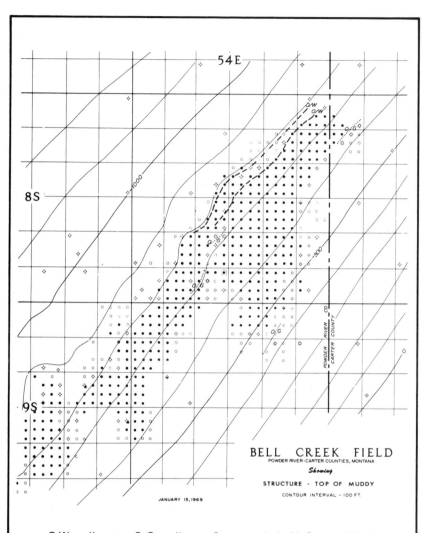

BELL CREEK FIELD
POWDER RIVER-CARTER COUNTIES, MONTANA
Showing
STRUCTURE - TOP OF MUDDY
CONTOUR INTERVAL - 100 FT.

JANUARY 15, 1969

O-W = oil-water; O-G = oil-gas. *Courtesy A.A. McGregor, Charles A. Biggs, and Samuel Gary.*

Figure 35

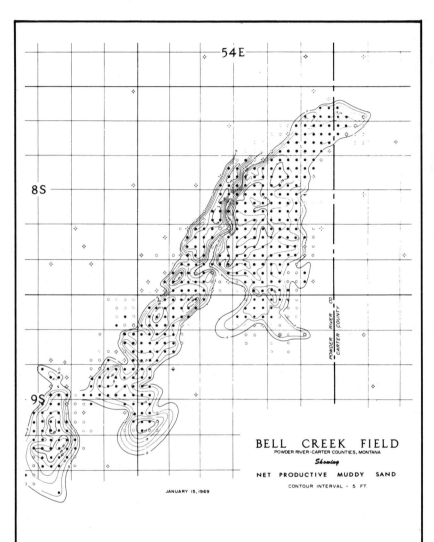

BELL CREEK FIELD
POWDER RIVER-CARTER COUNTIES, MONTANA
Showing
NET PRODUCTIVE MUDDY SAND
CONTOUR INTERVAL - 5 FT.

JANUARY 15, 1969

Courtesy A.A. McGregor, Charles A. Biggs, and Samuel Gary.

Figure 36

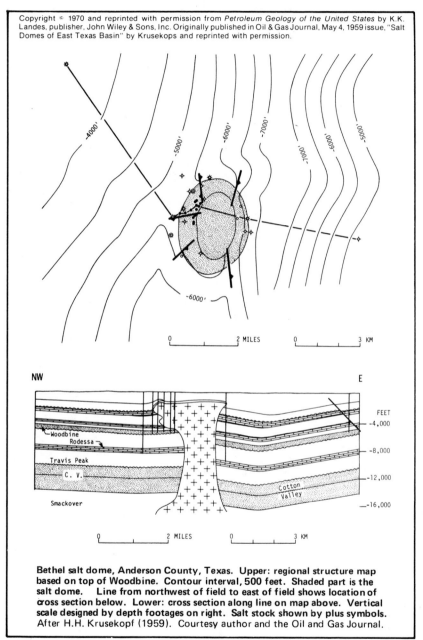

Bethel salt dome, Anderson County, Texas. Upper: regional structure map based on top of Woodbine. Contour interval, 500 feet. Shaded part is the salt dome. Line from northwest of field to east of field shows location of cross section below. Lower: cross section along line on map above. Vertical scale designed by depth footages on right. Salt stock shown by plus symbols. After H.H. Krusekopf (1959). Courtesy author and the Oil and Gas Journal.

Figure 37

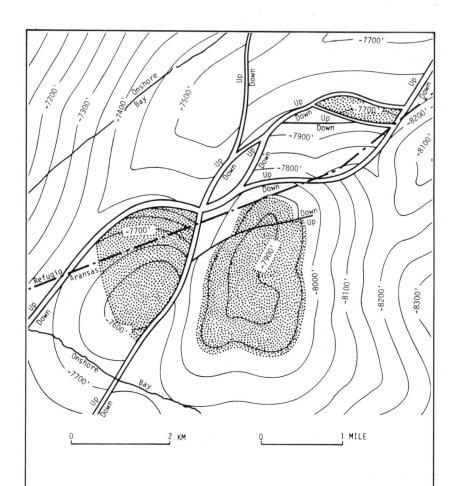

South Copano Bay gas field, Refugio County, Texas. Structure contour map based on top of Melbourne Sand (K-2 reservoir). Interval, 50 feet. Stippled areas show the larger separately trapped gas accumulations in this offshore field. Courtesy Leonard C. Bryant (1961) and the Gulf Coast Association of Geological Societies.

Figure 38

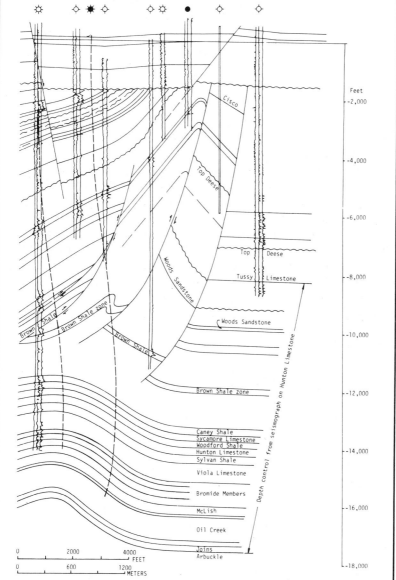

Carter-Knox oil field, Grady and Stephens Counties, Oklahoma. Southwest-northeast cross section. After Harold J. Reedy and Howard A. Sykes (1962). Courtesy of authors (per Harold J. Reedy) and the Tulsa Geological Society Digest.

Figure 39

65

portion and major thrust faulting is indicated. This is an example of an electric log cross section, although the logs have been much reduced for presentation purposes.

Another thrust-faulted field, Figure 40, is the Circle Ridge Field in Wyoming. The major accumulation is on the overthrust block but, as shown on the cross section, there is also subthrust production, and this is not an unusual situation.

Figure 41 illustrates the traps formed by rollover of the beds during contemporaneous faulting. This phenomenon is usually thought to be limited to the Gulf Coast, but there is some evidence that contemporaneous faulting is more widespread.

Finally, Figure 42, illustrating the Bisti Oil Field in New Mexico, shows the regional nature of stratigraphic traps. These sand bars are relatively thin and difficult to map, but they stretch over many townships and can be an attractive exploration target.

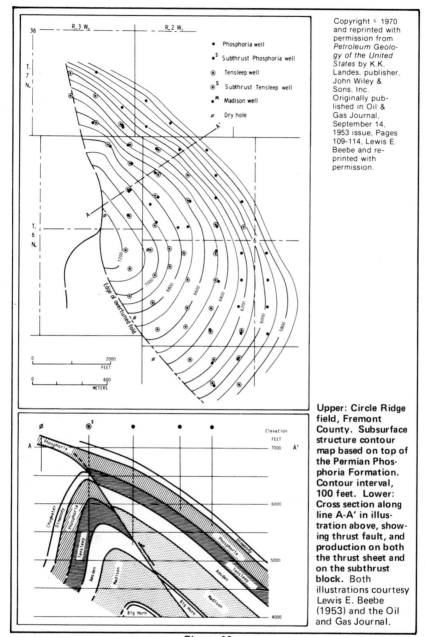

Upper: Circle Ridge field, Fremont County. Subsurface structure contour map based on top of the Permian Phosphoria Formation. Contour interval, 100 feet. **Lower: Cross section along line A-A'** in illustration above, showing thrust fault, and production on both the thrust sheet and on the subthrust block. Both illustrations courtesy Lewis E. Beebe (1953) and the Oil and Gas Journal.

Figure 40

67

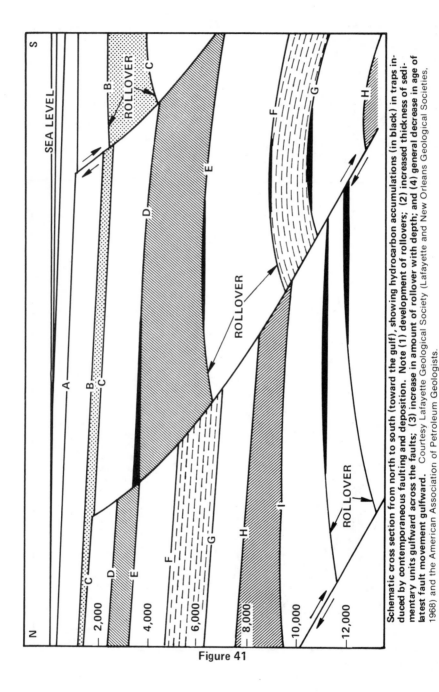

Schematic cross section from north to south (toward the gulf), showing hydrocarbon accumulations (in black) in traps induced by contemporaneous faulting and deposition. Note (1) development of rollovers; (2) increased thickness of sedimentary units gulfward across the faults; (3) increase in amount of rollover with depth; and (4) general decrease in age of latest fault movement gulfward. Courtesy Lafayette Geological Society (Lafayette and New Orleans Geological Societies, 1968) and the American Association of Petroleum Geologists.

Figure 41

68

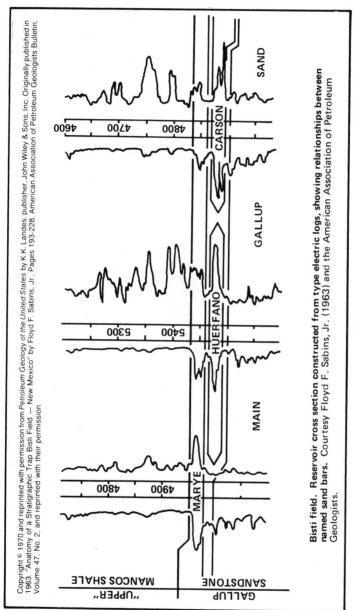

Bisti field. Reservoir cross section constructed from type electric logs, showing relationships between named sand bars. Courtesy Floyd F. Sabins, Jr. (1963) and the American Association of Petroleum Geologists.

Figure 42

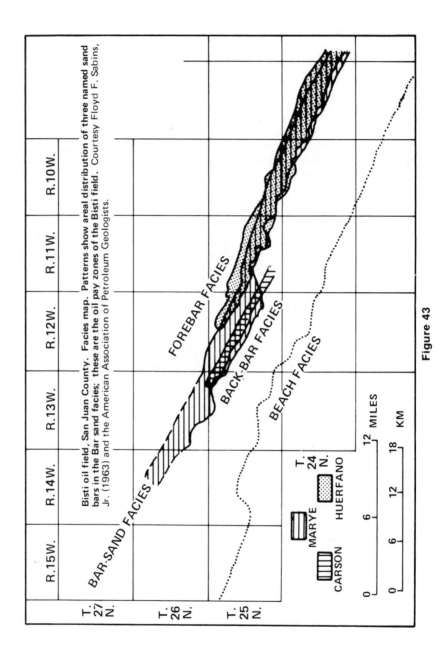

Bisti oil field, San Juan County. Facies map. Patterns show areal distribution of three named sand bars in the Bar sand facies; these are the oil pay zones of the Bisti field. Courtesy Floyd F. Sabins, Jr. (1963) and the American Association of Petroleum Geologists.

Figure 43

Chapter Three

EXPLORATION METHODS

§ 3.01 Geological Field Work

The original, and most basic, method for predicting what attitude the rocks will have in the subsurface, is to map the attitudes of the same rocks on the surface. Geological field work consists mainly of mapping the strike and dip of various formations. This is done with a compass and alidade, the basic instruments of the field geologist. Samples of the rocks also are taken and later examined under a microscope and described. Thicknesses of the formations are measured with a steel tape.

All of this information is transferred to a map and to written descriptions. In the early days of petroleum exploration and the anticlinal theory, surface geology, well mapped, was adequate to show the anticlinal features which might have oil or gas in them. The surface expressions of potential hydrocarbon traps have long since been drilled. "The easy ones have been found." So we are forced to look deeper at rocks which have no relationship to those on the surface. Now, in the mature areas of oil development, we do not do much surface geological measurement; we concentrate on subsurface data instead. But in new areas of exploration — new frontiers — surface geology is always one of the first methods used.

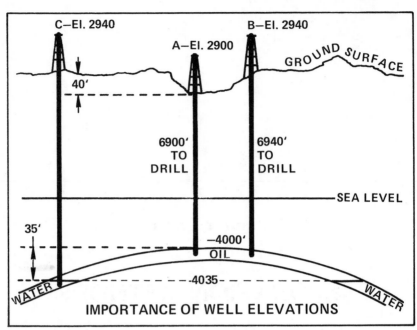

Figure 44-a

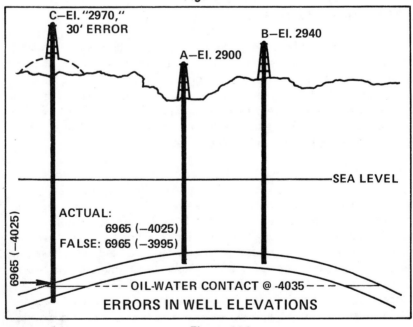

Figure 44-b

§3.02 Importance of Well Elevation

There are generally three elevations, relative to sea level, reported on each well: the ground elevation (GL), which is the permanent one, the derrick floor (DF) elevation, and the kelly-bushing (KB) elevation. Logs are usually run with the kelly bushing as the reference datum.

Whether or not the elevation needs to be very accurate depends on the surface relief and/or the subsurface relief, as shown by the following illustrations:

In Figure 44-a, the actual ground elevations are, for A: 2,900 feet; for B: 2,940 feet; and for C: 2,940 feet above sea level. (Sea level is the common denominator, or "datum" to which we tie all structural relationships.)

The formation to which A and B have been drilled is at the same depth *relative to sea level* in each well. But well B was on a hill and therefore will have to drill 40 more feet than well A. In high relief areas this difference could be 400 feet. So, predicting the drilling depth for each formation is dependent on the elevation.

Suppose that well C was planned to drill just the top five feet of the reservoir rock because to drill deeper would mean drilling into water. An error in calculating the elevation could mean that the well would be drilled too deep because of a false impression that the well was "running high" unexpectedly. The falsely optimistic situation is shown in Figure 44-b.

Based on previous wells drilled on the structure, the oil-water contact is known to be at minus 4,035. Well C is expected (correctly) to encounter the producing zone at minus 4,025, and is programmed to cut five feet and stop. However, because of an erroneous elevation (30 feet too high), too much section is cut and the well encounters too much water to be a successful completion. The total amount of hole to be cut is 6,965 feet to the top of the producing zone which is, in actuality, at minus 4,025 (6,965 minus the true elevation of 2,940). However, when the 6,965 feet have been cut and the false elevation is subtracted (6,965 - 2,970 = 3,995), the result is a report that the well is "running high" and a probable decision to drill the whole "pay" section.

In general, in areas of low relief, in either surface or subsurface or both, a relatively small error of a few feet in the elevation can be significant. Western Kansas would be an example. In such areas, maps are sometimes made with a contour interval of 10 feet or 20 feet. On a map with a contour interval of 10 feet, a three-foot error in the elevation could result in a significant anomaly, high or low, which can cost a lot of money if acted upon.

Usually, however, the geologists working such an area are acutely aware of this problem and can handle it by increasing the contour interval to something like 30 feet or more. And, of course, there are ways to double-check the elevations by consulting topographic maps published by the U.S.G.S. and adding the height of the derrick floor or kelly bushing to the ground elevation.

In areas of high relief, either in the surface or subsurface, as large an error as 40 or 50 feet (which should be rare) probably would not make much difference, especially in wildcatting. In development drilling it can become important, but a small error can be tolerated.

In the Midcontinent, for example, when the elevation is subtracted from the drilling depth, the result is below sea level and is called the "subsea" elevation. For instance, elevations in central Oklahoma are approximately 1,200 feet above sea level. Most drilling depths of recent wells there are on the order of 12,000 to 15,000 feet so that when 1,200 is subtracted, the relative elevation is well below sea level. On maps, we use a minus sign to show this.

In the Rocky Mountains, for another example, many wells are drilled high up in the mountains — say at an elevation of 10,000 feet above sea level. The situation is exactly the same as far as the reference datum — sea level — is concerned. If an important horizon occurs at 3,500 feet, one still subtracts the surface elevation, in this case 10,000 feet. The answer is 6,500 above sea level and is shown with a plus sign (or no sign) on a map.

§ 3.03 Bore Hole Devices

Bore hole measuring devices, and the recordings or logs produced from them, are dealt with in some detail in a separate chapter on Logging

Methods. Here the emphasis will be on the uses of logs in exploration only.

So-called electrical logs (also called mechanical logs) are measurements of some physical characteristic of the rocks which were cut when the well was drilled, or of the fluids in those rocks. The most widely used device is a combination one measuring, (1) the natural electrical current (spontaneous potential or S.P.) present in the bore hole, and (2) measuring the response of the rocks to an emitted electrical current. The latter is a measurement of the conductivity of the rocks. The opposite or reciprocal of conductivity is resistivity. Some form of these two logs (S.P. and resistivity) has been used from the earliest days of logging and still is one of the most useful for correlation.

For exploration purposes, a major use of logs is for correlation from well to well. Accurate correlation is a must in building a prospect. A mis-correlation is a bad geological error, but it can happen quite frequently if the well control is sparse. Rocks are simply not laid down like a layer-cake, contrary to earlier — and still popular — ideas.

The other major use for logs in exploration is for quantitative interpretation of the fluids in a given zone, as well as the porosity and permeability. Wildcat locations can be predicated on well-log analysis which may suggest drilling either up dip or down dip from a dry hole which had shows. In the chapter on cross sections there is a discussion of the uses of electric logs in depicting the presence of a trap.

§ 3.04 Sample Logs

In the early days of the oil business there were no mechanical devices to run in the bore hole to obtain a graphic record of the formations which were penetrated by the drill-bit. Therefore, the rocks themselves, ground up by the bit but still recognizable under a microscope, became the primary log of a bore-hole. Samples of the rock, usually called cuttings, are collected at intervals of five feet, 10 feet, or other specified intervals. They must be washed free of the mud that carried them to the surface from the depth where they were drilled. Then, they are dried and put into paper envelopes for later examination.

"Running samples" is slowly becoming, if not a lost art, certainly a restricted one, practiced by a few old-timers and the specialists who work on mud-logging trucks. There are still some advantages to looking at the samples rather than depending solely on an electric log. For instance, the color of the rocks can be a useful correlation tool, but no electric log can report that. Oil staining on the rock can be seen and can be evaluated to some extent, based on experience. Is the stain uniformly distributed or spotted here and there on the rock? Is it heavy, almost asphaltic or light, high gravity oil?

In the rock lithologies, also, there can be subtle but significant inclusions which may be the key to interpretations of environment. For example, if a sandstone carries inclusions of carbonaceous material such as wood fibers and plant material along with considerable clay, the interpretation would be that the sandstone was adjacent to or part of a lagoonal deposit. The explorationist would look for cleaner sand away from the lagoonal area.

§ 3.05 Driller's Logs
Because we have sophisticated logging methods now, the driller's log is no longer needed on modern wells. But they are very useful, if available, on very old wells which may never have had any mechanical logs run. The driller would record the kind of rock he cut through in a certain interval with phrases such as "shale and shells" or "blue gumbo," adding comments about shows of gas, or oil and water when they occurred.

§ 3.06 Drilling Time
The speed with which the bit penetrates the rock is not solely of economic importance. It is also a clue to the kind of rock being drilled. A common phrase is that a sand "drilled off", meaning that it was fairly soft and could be cut quickly. This means that the sand probably has good porosity and permeability. Another use for drilling time is to pick the top of a zone which is to be drill-stem tested. There is often considerable difference between the penetration rates of sand and shale, and so the boundary

between the two can be picked quite accurately before any mechanical logs have been run in the hole.

Drilling time is recorded in minutes per foot. Originally, the drillers kept this record by hand, by simply checking their watches every few feet and writing it down. This method was not always very accurate. Now there are automatic recording devices.

§ 3.07 Drill Stem Tests

A drill stem test is not a logging device but a procedure to try to determine what fluids are in a given formation. A perforated pipe with an expandable packer on it is attached to the bottom of the drill string. When the packer is opposite the place in the hole which has been picked as the "packer point," the packer is expanded to seal off the formations uphole. Then the tool is opened opposite the formation to be tested. Any fluids and/or gas in the formation should flow into the testing tool and the amounts can be measured when the equipment is pulled out of the hole. Bottom-hole pressures are measured while the test is carried on, and all of this information is interpreted to try to determine whether the zone is "pay" or not.

Unfortunately, there are many inconclusive drill-stem tests. But when the fluid recovery is definitely oil or gas or salt water, then there is a firm basis for decisions such as whether to drill deeper and whether to set production casing or not.

§ 3.08 Surface Measurements

Beside the measurements made with compass and alidade, described under the heading "Geological Field Work," there are some additional measurements which can be made on the surface in the search for oil in the subsurface. Some of them are quite sophisticated, others not. All of them have some merit, but none is regarded as the answer to an explorationist's prayer.

§ **3.08(a) Soil Sampling.** This technique assumes that some hydrocarbons seep upward from a deep-seated accumulation and can be found in minute quantities in the overlying soil. The soil samples are analyzed in the laboratory for the presence of hydrocarbons. There are believers and non-believers in this method.

§ **3.08(b) Near-Surface Geophysics.** A few attempts have been made to use near-surface recording techniques to explore for anomalies. Two examples are electrical resistivity logging and gamma-ray logging. The depth of investigation of these methods ranges down to about 2,500 feet. These methods have not been very successful.

It should be noted that regular exploration geophysics, such as gravity, magnetics, seismic and magneto-tellurics measurements, while conducted at the earth's surface, are recording information from considerable depth and therefore should not be regarded as surface tools.

§ 3.09 Maps

§ **3.09(a) Contour Lines.** The most common map used in support of a prospect is the structure map. The lines on such a map are lines connecting points (real or estimated) of equal structural position with respect to sea level. The resulting shape looks like rather smooth surface topography. The lines are the contours of a subsurface feature.

The lines on an isopach map connect points (real or estimated) of equal thickness. Again, a shape emerges which can be related to some subsurface feature, such as a river channel.

There are certain rules for drawing contour lines which are universally applicable, no matter whether the information being "contoured" is structural position, thickness, initial potential, pressure, temperature, or whatever.

Before listing these rules, for those who wish to try contouring for themselves, there are some generalizations which can be made which should be useful to the observer of others' maps. For one thing, the general appearance of the map should be one of fairly smooth, curving, and regularly spaced lines between control points. The map should present some clearly defined shapes or anomalies, if the objective is supposed to be structurally high. All data points should appear on the map so the observer can see where the control is good and where it is poor.

Mapping can be an exploration method in itself. Nothing is more satisfying to a geologist than, after having "posted" his data on a map, to draw contour lines and find an anomaly. Here is the usual procedure:

If it is determined (or guessed) that a structure map of a certain area would be desirable then the data must be first collected. This is usually done by selecting some particular zone on an electric log which also occurs on most, if not all, the other logs in the area. The accurate correlation of this zone in all well logs is vitally important. As stated before, a mis-correlation is a bad geological error but it certainly can happen to the most skilled geologist.

The next procedure is to determine the bore-hole depth of the horizon to be mapped, subtract the elevation, and "post" it, or enter it beside the well-symbol on the map, as shown below:

-2952

Sometimes more than one piece of information is noted for each well. For instance, if the thickness of a zone is to be mapped also (isopach map), then it is customary to post the data like this:

-2952 (14)

When this information is drafted on a base map two reproducible copies can be made, one for contouring the structural information and one for contouring the thickness information. Blue-line prints can then be made from these reproducibles.

Sometimes maps are used for data storage and rarely contoured in that form. Such a map will have well-symbols with "data-trains" beside them and look like this:

Hx-1908
DS-2572
BP/V-3102
Simp-3791
Arb-4925
TD 6227 (-5025)

The elevation of the well (1,202) is shown either separately from the data-train, or at the very top of it. The sub-sea tops of important formations or zones then are listed with abbreviated names, and occasional thicknesses are also shown, depending on their importance to the geologist. The Total Depth (TD) is usually given as actual drilling depth plus the sub-sea value.

A Base Map is a map on which only well-symbols (well-spots) are shown. These are the usual symbols for dry holes, oil wells, gas wells, etc. A list of these is given in the Appendix.

Often, a map showing well-symbols and the accompanying data train is also called a Base Map, probably because it contains the basic data a geologist works with. It is common to reproduce portions of such a map and then draw structure contours on one of the horizons listed. This is sometimes a first approach to developing a prospect.

To summarize the procedures which result in a single map of a prospect (usually there are many maps), the following steps are taken:

1. Correlating of well data.
2. Selecting of mappable zone.
3. Reading of drilling depth and subtraction of elevation to get the elevation relative to sea level. Refer to Chapter 3 § **3.02** for the importance of well elevations.
4. Posting of data on map beside well symbol.
5. Contouring.

Here are some rules for contouring:

§ 3.09(b) Rules:

General

Contour lines connect points of equal value.
Contours never cross one another.
Contours should not dangle, unconnected.
Choice of contour interval depends on, among other things: density of control points, scale of map, amount of possible error in data, and type of feature expected. The choice really boils down to the question of how many lines need to be drawn to make a useful map. Equally spaced contours mean that the rate of change (of slope or dip, thickening or thinning) is constant.
When contour values change from increasing numbers to decreasing, or vice-versa, the highest or lowest contour must be repeated, unless a perfectly "V" shape is expected.

Structure and Topographic Maps

The contour interval is a vertical measurement. It should remain constant on any map unless it is clearly stated in the legend that the interval changes over the map.
Closely spaced contours indicate steep dip (slope) or faulting.
Contours wide apart indicate gentle dip (low relief).
Where drainage is involved, as on a topographic map, contours always bend upstream.

Isopach Maps

Since the contour interval is usually small, data must be accurate. Alternatively, if error is expected, contour interval must be large enough to tolerate the error and still be meaningful.
Closely spaced contours indicate rapid thinning or thickening.
Evenly spaced contours indicate uniform rate of thinning or thickening.

§ 3.10 Ideas

When all is said and done, ideas are what count in the oil exploration business. While it is possible to assemble a lot of data, make a structure contour map and have an anomaly — a structure — pop out at you when you didn't know it was there, it is not common these days. More usually, one has an idea ahead of time — a preconceived idea — and then tries to find out whether the idea has any merit or any support. The notion of a preconceived idea has a slightly unpleasant connotation or rigidity and hard-headedness which I don't think is appropriate when used in oil exploration. What happens is more like this:

After working in an area for some time, a geologist gets a feel for the various geological aspects which are normal. Then he looks for the abnormal, the anomalous; anomalies are his business. An unexpected thinning of certain rocks, a missing section caused by a fault not previously identified, an electric log curve reading a bit higher than usual — these are all anomalies which could lead to the development of a new prospect.

On identifying an anomaly, the geologist tries to interpret what it means. Could the new fault form a trap for oil and gas? Does the thinning mean that a deeper structure exists? Does the curve on the electric log mean that a channel sand may be developing nearby? These are the questions a geologist asks when he sees something anomalous. In pursuit of the answers, there are many dead-end trails, as one might expect. It is so easy to "wipe out" a potential prospect and so hard to find really strong support for one. Most prospects have some weaknesses, but if the basic idea is geologically sound and there is no damning evidence against it, the idea is worth testing.

There is quite a similarity to detective work, it seems to me. There is a lot of basic drudgery in gathering information (evidence), a lot of false leads, worry about having overlooked some detail, and when all available information is at hand, general dissatisfaction with the support for your prospect (case).

Nevertheless, the rewards can be worthwhile. Intellectually, it is satisfying to see one's idea proved out when it was based on rather sketchy bits of information. In some cases it can be financially rewarding also.

As in any field, there are journeyman geologists and intellectual geologists and the great majority which falls in between. Throughout these ranks, the prospect finder is the geologist with ideas.

Chapter Four

LOGGING METHODS AND APPLICATIONS

By E.W. "Bill" Sengel

§ 4.01 Introduction

Well logs are continuous recordings versus depth of measurements made in a bore hole. These measurements of one or more petrophysical characteristics (described later) of the rocks penetrated by the bit are made with specially designed instruments lowered into the hole on a wire line.

It has been the practice for many years for operators to run one or more logs in practically every well drilled, whether a dry hole or a producer. Consequently, there are many thousands of logs available in every oil and gas province.

The reasons for running logs are many: they are a permanent continuous record of all formations cut by the well bore; they are a source of the information desired from the drilling of the hole, such as information used by geologists in their search for oil or gas; and, of great importance, they provide data used in the evaluation of producing formations.

It is not the intent nor purpose here to make one an expert in the field of well logging, but to expose him to the techniques. Should an interest be created, there are great quantities of publications available through the various service companies that will allow in-depth studies to be made. The field is very complex and there should not be attempts to oversimplify it. Full knowledge of the procedures and effects of variables is extremely important. It is a case of a little knowledge can be dangerous, improper results can be very expensive.

§ 4.02 Information Desired from the Wellbore

When a hole is drilled in the ground there are several things the operator hopes to learn. Of prime importance is the identification of possible producing zones. Of the many rock layers cut by the bit, only some are porous. The rest are shales, tight sandstones, hard carbonate sections, etc. Of the porous zones, only some of them have the proper conditions for hydrocarbon accumulation. They must have some sort of trapping mechanism, and be in a position that hydrocarbons migrating from a source bed can get into them.

Rocks can be identified by their chemical and physical characteristics. These are generally color, texture, hardness, and fossil content. In the case of oil or gas bearing rocks there may also be stain and/or odor. These are direct measurements and drilling samples or cores are necessary.

There are also petrophysical characteristics that can help identify the rocks. These are resistivity, spontaneous potential, elasticity, electron density, radioactivity and nuclear. They are, or can be, indirect measurements made with different logging devices lowered into the borehole.

If possible producing zones are located they must be evaluated.

The pores (openings) in the rock provide the storage space for the oil or gas, so it is neceesary to determine the porosity of the zone to see if there is enough to make a commercial reservoir.

Generally, the rocks are deposited in water so that in the beginning the porosity is filled with water. The hydrocarbons migrate into them later displacing the original water. It is necessary to determine whether enough water has been displaced so that only hydrocarbons will be produced.

Another necessary bit of information is whether or not there is enough permeability to allow the zone to produce in sufficient quantities.

Then, of course, there is need to know the thickness of the zone. Thickness related to porosity and hydrocarbon saturation helps determine the economic value of the reservoir.

Correlation of various rocks is also desired from the well. The relative position of particular zones with depth is used in subsurface mapping for exploratory or development work.

Also necessary is accurate depth control. Completion techniques (perforations, plugs, packers, etc.) must be correctly placed, especially when thin zones or water contacts are involved.

§ 4.03 Electrical Characteristics of Rocks

There are two basic electrical characteristics: resistivity and spontaneous potential.

Resistivity is the ability of a substance to impede the flow of electricity through it.

Generally, sedimentary rocks are made up of various minerals such as quartz, silicates, carbonates, etc., and when they are dry they are non-conductors of electricity. Therefore, they have high resistivity. Hydrocarbons also are non-conductors, so they have high resistivity.

There are conductive solids in nature also, such as pyrite, magnetite, vermiculite, illite, glauconite, etc. These may be found in formations in sufficient quantity to effect the formation resistivity. They can create problems when analyzing the formation from logs.

Since the rocks were deposited in water and the water never is displaced completely by hydrocarbons, it will have a definite effect on the resistivity of the formation. Water is a conductor of electricity, therefore it has low resistivity. Its resistivity is inversely proportional to its salinity, at a given temperature. High salinity water has low resistivity, and as salinity decreases the resistivity increases. Temperature will effect the water resistivity also. For a given salinity the resistivity will decrease as the temperature increases. Water found in most subsurface formations has high salinity; therefore, it has low resistivity.

Formation water resistivity is very important in log analysis work. For several years there has been much effort spent in obtaining good values from produced water from many zones. This information has been catalogued so that now there is very much information available concerning formation water resistivities in most areas.

This is a good time to digress and make some remarks pertinent to symbols and formulas used in working with well logs. Figure 45 is a list of basic symbols. There are many more in use, but this list will suffice until one gets more involved in log work. Mathematics will be kept at a minimum in this article; however, there are some basic relationships that are necessary. They will be introduced at their appropriate places in the text.

To illustrate the use of symbols, the previous paragraph refers to the resistivity of the formation water. The symbol for this is Rw. The "R" indicates resistivity, and the subscript "w" indicates that water is the substance measured. Other examples: Rm is the resistivity of the drilling mud, Rmf is the resistivity of the mud filtrate, and Rmc is the resistivity of the mud cake.

Now back to discussing resistivity. The resistivity of a formation can be expressed by the following relationship:

$$\text{Resistivity (R)} = \text{Rw} / \emptyset^m \times \text{Sw}^n$$

$$\text{Rw} = \text{formation water resistivity}$$
$$\emptyset = \text{formation porosity}$$
$$m = \text{formation cementation factor}$$
$$\text{Sw} = \text{formation water saturation}$$
$$n = \text{saturation exponent (generally 2)}$$

This indicates that the formation resistivity is directly proportional to the resistivity of the formation water (Rw) and inversely proportional to the amount of porosity and to how much water is in the porosity. If a given formation has no porosity ($\emptyset$), there is no water involved so the formation has extremely high resistivity. As the porosity ($\emptyset$) increases and water saturation (Sw) is 100 percent the formation resistivity decreases. If the formation has a given porosity ($\emptyset$) and water saturation (Sw) decreases

SCHLUMBERGER WELL SURVEYING CORP.

SYMBOLS

AM	—Normal Curve Spacing
AO	—Lateral Curve Spacing
BHT	—Bottom Hole Temperature in °F
d	—Diameter of Hole
Di	—Average Diameter of Invaded Zone
e	—Bed Thickness in Feet
t_{me}	—Mud Cake Thickness
R_m	—Resistivity of the Mud
R_{mf}	—Resistivity of the Mud Filtrate
R_{mc}	—Resistivity of the Mud Cake
R_w	—Resistivity of the Formation Water
R_{wa}	—Apparent Resistivity of the Formation Water in Shaly Sands
R_z	—Resistivity of the Mixture of the Electrolytes *(It can be defined for a given zone—invaded or noninvaded)*
R_t	—Resistivity of the Formation—Uncontaminated Zone
R_o	—Resistivity of the Formation when 100% Water
R_i	—Resistivity of the Invaded Zone
R_{xo}	—Resistivity of the Flushed Zone *(Close to Bore Hole)*
R_s	—Resistity of the Surrounding Beds
$R_{16''}, R_{64''}, R_{18'8''},$ $R_{1''x1''}, \& R_{2''}$	—Apparent Resistivity of 16" Normal, 64" Normal, 18'8" Lateral, 1"x1" Microinverse, and 2" Micronormal, respectively
$R_{IL6},$ or $R_{ILd},$ R_{IL5} and R_{ILm}	—Apparent resistivity of 6FF40, 5FF40, and Medium Induction, respectively
F	—Formation Resistivity Factor
F_a	—Apparent Formation Resistivity Factor
ϕ	—Effective Porosity in Per Cent
S_w	—Water Saturation, Per Cent of Pore Space in Uncontaminated Zone
S_i	—Water Saturation, as above, in Invaded Zone
S_{xo}	—Water Saturation, as above, in Flushed Zone
ROS	—Residual Oil Saturation as Per Cent of Pore Space; = $(1-S_{xo})$
K	—Coefficient in the SP Formula
SSP	—Static Spontaneous Potential—The Maximum Possible for a Particular R_{mf}/R_w
PSP	—Pseudostatic Spontaneous Potential—The SP Found in a Thick Shaly Sand
α	—SP Reduction Factor — PSP/SSP
k	—Permeability in millidarcies
Δt	—Sonic Transit Time in miscroseconds per foot
ρ_B	—Bulk density in grams per cubic centimeter

FORMULAS

Archie's Saturation Formula:	$S_w{}^n = F R_w/R_t$; $S_w = \sqrt[n]{R_o/R_t}$. (n usually taken as 2)
Formation Resistivity Factor:	$F = R_o/R_w$; $F_a = R_{xo}/R_{mf}$.
Formation Resistivity Factor vs. Porosity:	Humble: $F = 0.62/\phi^{2.15}$; Archie: $F = 1/\phi^m$.
SP Formulas:	$SSP = -K \log_{10} R_{mf}/(R_w)e$; $PSP = -K \log_{10} R_{mf}/R_{wa}$.
Sonic Time-Average Formula:	$\Delta t = \dfrac{\phi}{V_{fluid}} + \dfrac{(1-\phi)}{V_{matrix}}$
Formation Density Formula:	$\rho_B = \phi \, \rho_{Fluid} + (1-\phi) \, \rho_{Grain}$

LOG INTERPRETATION CHARTS
REFERENCE SHEET

Figure 45

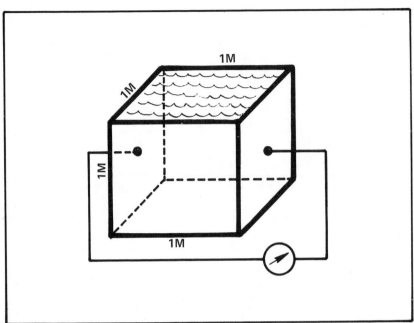

Figure 46

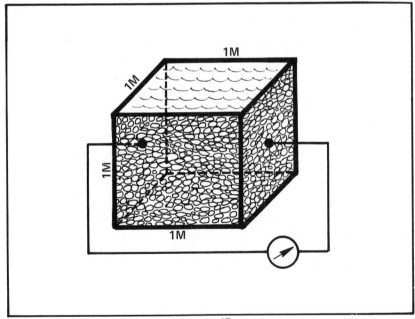

Figure 47

from 100 percent, then formation resistivity will increase. For a given porosity (Ø) and water saturation (Sw), the formation resistivity will be lower with high salinity water (low Rw) than it will be with a low salinity water (high Rw).

It is important to realize that formation resistivity is measured through the water contained in the porous spaces. Therefore, it is affected by the water resistivity (Rw) and the rock itself. The resistivity contributed by the rock is called formation resistivity factor, commonly shortened to formation factor (F). An understanding of formation factor (F) is imperative. In Figure 46 we have an insulated square container one meter on all sides. It is level full with water. The resistivity is measured between two opposite faces and gives a value, for example, equal to 1. Since there is 100 percent porosity (Ø) and 100 percent water saturation (Sw), the resistivity measured is the water resistivity (Rw).

The container is a square cubic meter so the water resistivity (Rw) is 1 ohm-meter. The ohm-meter is the unit of resistivity used in well logging. It is always for a square cubic meter and is expressed on the log headings as resistivity per M²M (meter square meter).

In Figure 47 we have the same container only now sand has been added. It is still full of the same water.

The resistivity measured between the two opposite faces has increased to, let's say 10 ohm-meters. There is still 100 percent water saturation (Sw) with the same water resistivity (Rw) of 1. The only reason the measured resistivity is higher is because formation has been added so the porosity (Ø) is now less than 100 percent. The additional resistivity is due to the formation factor (F). In this case it is 10.

Figure 48 has the container with the same situation as in Figure 47, only now the sand grains are cemented together with some sort of cementing material.

Now the resistivity measured between the two opposite faces will be higher, for example 15 ohm-meters. The water saturation (Sw) is still 100 percent and water resistivity (Rw) is still 1 ohm-meter. The porosity (Ø) is still the same as in Figure 47. The only difference is that the sand grains are cemented. Formation factor (F) is now 15. Formation factor (F) can be defined as the number by which water resistivity (Rw) can be multiplied to determine the resistivity of a formation at 100 percent water saturation (Sw). As indicated by the previous examples, it is related to porosity and cementation. A formula for it is: $F = 1 / \emptyset^{m}$. The exponent "m" is the cementation factor and may vary from 1.3 in soft unconsolidated sandstones to as high as 3 in some carbonates.

Formation resistivity is measured with specially designed instruments, the type depends on the conditions in the hole. Wells are generally drilled

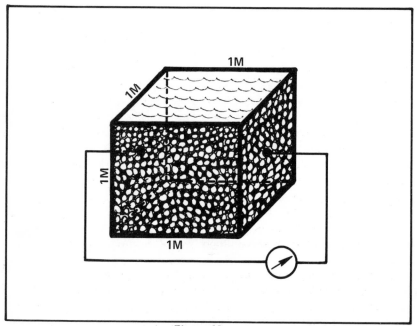

Figure 48

with rotary tools under one of the following conditions: water-based mud, with either fresh or salty make-up water, oil-based muds, or gas (air) filled.

Logs used to measure resistivity for each hole condition are:

1. Fresh mud (mud filtrate less saline than formation water).
   (a) Electrical Log (old style).
   (b) Induction — Electrical Log.
   (c) Dual Induction — Laterolog 8
2. Salty mud (mud filtrate about equal salinity as formation water).
   (a) Laterolog.
   (b) Dual Laterolog.
   (c) Dual Induction — Laterolog 8 (in certain circumstances).
3. Oil base muds.
   (a) Induction type log.
4. Gas (air) filled.
   (a) Induction type log.

Resistivity measuring curves are recorded in the two tracks to the right of the depth column on the log.

The purpose of resistivity measurements is to determine the true resistivity of the formation, overcoming the effects of the borehole and invasion by the mud filtrate.

Let us turn our attention to Figure 49. This is a schematic presentation of what happens when a porous, permeable formation is penetrated by the drill bit, using a water-based mud system.

The porous permeable zone is the shaded interval. As the bit penetrates this zone filtrate from the mud begins to enter the porous spaces, flushing formation fluids ahead of it. As this action is taking place, solids in the mud are plastering the wall of the hole forming mud cake. The mud characteristics control these actions. The action of the filtrate entering the formation is called invasion. The invading fluid will flush out all the movable formation fluids for some small distance away from the wall of the hole and create the "flushed zone." Invasion may continue but flushing will become less as it gets farther away from the wall of the hole. At some point there is no more invasion and beyond that point the formation contains only the original fluids. The zone between the flushed zone and the uninvaded or uncontaminated zone is the "invaded zone." The invaded zone can have many conditions depending on time, mud characteristics, rock characteristics, original formation fluids, etc. In any event, it is a problem to be overcome in obtaining true resistivity (Rt). The resistivity of the flushed zone is known as Rxo. The fluid in the flushed zone is primarily mud filtrate (Rmf) with some of the original fluids left. The saturation of mud filtrate in the flushed zone is Sxo. The invaded zone resistivity is called Ri. The fluid in the invaded zone is a mixture of mud filtrate and formation fluids and is called Rz. The saturation of fluid mixture in the invaded zone is Si. Beyond the invaded zone is the uninvaded or uncontaminated part of the formation. The resistivity of the uninvaded zone is called Rt. The fluids in the uninvaded zone will be formation water (Rw) and hopefully some oil or gas. The saturation of water in the uninvaded zone is Sw.

The adjacent beds are generally shales or impervious beds that are not invaded by the mud filtrate.

The use of true resistivity (Rt) will be discussed later when the analysis of logs is developed.

The spontaneous potential (S.P.) is a measurement of potential variations in the borehole caused primarily by the interaction of the mud filtrate in the hole and the formation water. There are other factors that help cause the S.P., but the ratio of the mud filtrate resistivity (Rmf) to the formation water resistivity (Rw) is the most important.

The S.P. curve is useful in detecting porous permeable beds and their boundaries, indicating the shaliness of a formation, determining a value for formation water resistivity (Rw) when no other is available, and correlation.

The spontaneous potential (S.P.) is recorded in the single track of the log left of the depth column. It is adjusted by the logging engineer so that the shale readings are near the right side of the track (near the depth column). Deflections to the left from the shale line are considered to be "negative" and those to the right from the shale line are considered to be "positive."

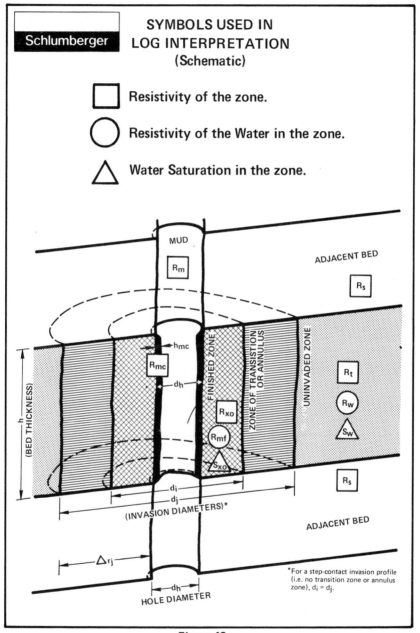

Figure 49

The ratio of the Rmf to the Rw controls the polarity of the S.P. curve. If the ratio is large the S.P. will be negative. This is the case where the mud filtrate is fresh and the formation water is salty. If the ratio is near to unity (1) there will be no S.P. This is the case where both the mud filtrate and the formation water are salty (they could both be fresh also). If the ratio is a fraction the S.P. will be positive. This is the case where the mud filtrate is saltier than the formation water. (See Figure 50.)

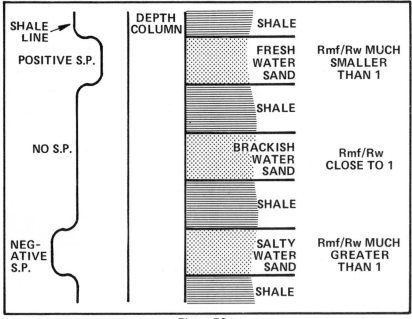

Figure 50

The S.P. responds to the presence of a porous permeable zone but the magnitude of the S.P. deflection has no relation to the amount of porosity or permeability.

Many things affect the S.P. curve. Among them are borehole diameter, invasion, bed resistivity, bed thickness and shaliness. All of them tend to decrease the S.P. deflection.

§ 4.04 Porosity

Reference has been made about porosity and its effect on both resistivity and the spontaneous potential. Knowledge of formation porosity is so important that several logging techniques have been designed to measure it. No attempt will be made to describe or explain the instrumentation of these logging devices. Should this knowledge be desired, there are papers on each available from the logging companies. Also, the interpretation techniques will be discussed later. The logs in general use to determine porosity are: wall resistivity devices (MicroLog, MiniLog, etc.); sound measuring devides (Sonic Log, Accoustic Log, etc.); density measuring devides (Density Log, Densilog, etc.); and hydrogen measuring devices (Neutron Log, Sidewall Neutron Porosity Log, Compensated Neutron Log, etc.).

The MicroLog was designed to detect permeable zones by indicating the presence of mud cake. This is a resistivity device that is mechanically pressed against the wall of the hole while the log is being run. Its measurements are confined to the flushed zone discussed earlier. Since determination of porosity is so important to formation evaluation, a mathematical solution of the MicroLog readings was developed to determine the formation factor (F). If F is known then porosity ($\emptyset$) can be approximated. There are many variables involved in the solution of the MicroLog which makes porosity determination subject to many inaccuracies.

The Sonic Log or Accoustic Log is a measurement of the time required for a sound wave to travel through one foot of formation. This is called interval transit time and is the reciprocal of the velocity of the sound wave.

The interval transit time for a given formation depends on its lithology and porosity. If the lithology is known the interval transit time can be useful in determining porosity. It can also be helpful to geophysicists in interpreting seismic records. Nearly all of the equipment in use at the present time is of the borehole compensated type (BHC). This practically eliminates effects from the hole such as tool position and hole size changes.

The sonic log may be affected by shaliness in the formation, gas and mineral mixtures which singly, or in combination, can cause erroneous porosity determination.

The density log is a measurement of the electron density in a formation, which varies with porosity. The density tool is mechanically pressed against the wall of the hole while logging. Present equipment is of the compensated type (FDC), which minimizes the effects of mud cake and hole roughness. The density log is less affected by formation shaliness but can be affected by gas and mineral mixtures.

The neutron log responds primarily to the presence of hydrogen, which is usually present in the liquids in the porous spaces. It can be affected by gas, shale, borehole size and formation mineral mixtures. There are several types of neutron logs which can be run for different hole conditions. Many improvements have been made in recent years to improve the quality of neutron measurements. These also have made them much more accurate. One of these is the Sidewall Neutron Porosity Log (SNP). This device is designed to be pressed against the wall of the hole while logging. This minimizes the borehole effect. It also utilizes higher energy neutrons which overcomes most problems caused by formation minerals and chemicals. Any corrections necessary are performed by the control equipment at the surface. The recorded curve is scaled on the log in linear porosity units. This log can only be run in uncased holes.

Another type neutron log is the Compensated Neutron Log (CNL). This is a mandrel type tool which uses dual spacings. The instrumentation is improved to the point that most all corrections necessary are performed by the control equipment at the surface. It is also scaled on the log in linear porosity. This log has the advantage of being able to run in either uncased or cased holes. It has a greater depth of investigation than the SNP. This has an advantage when used with other logs to detect gas.

The neutron logs can be run in combination with other logs such as the Gamma Ray, caliper log, compensated density log, sonic log and others.

The Gamma Ray log is a measurement of the natural radioactivity of the formations. Shales are highly radioactive while reservoir type rocks such as sandstones, limestones, and dolomites have low radioactivity. Thus the Gamma Ray can complement the S.P. curve when the S.P. is undependable or inoperative (Rmf = Rw, for example). It can help to determine shaliness of zones. It is also excellent for correlation of formations. It can be run in cased or uncased holes. The Gamma Ray log can be run in conjunction with many other logs such as the Dual Induction — Laterolog, Sonic Log, Density Log Neutron Log and Caliperlog.

§ 4.05 Evaluation from Well Logs

A modern "set" of well logs may be quite varied as to which logs are included. The selection of logs to be included will depend on many things: borehole conditions; types of formations encountered; whether or not information is available from other sources; and extent of evaluation desired, to name only a few. For example, there may be the rather uncomplicated situation where the well was drilled with fresh water mud, the formations are predominantly shales and sandstones, it is a development well with much reservoir data available from previous work, and the desired information is the presence or absence of the pay zone(s) and the quality. In this case, the logs selected will be a primary log, which is run from surface to total depth, and a porosity log, probably with a Gamma Ray and caliper. The primary log may be an Induction — Electrical Log, or a Dual Induction — Laterolog. There will be several resistivity curves and a spontaneous potential curve. This will provide information as to presence of zones, correlation and evaluation. The porosity log could be the Sonic Log, Compensated Density Log, or the Compensated Neutron Log, and each may be run with Gamma Ray and caliper. The Gamma Ray will provide information as to the shaliness of the zones and the caliper will provide information about hole conditions for completion work. The porosity log will provide information as to the storage space. The porosity and resistivity values from the primary log will permit evaluation of fluid content.

A primary log will always be selected because it provides a record of the hole. Beyond that the selection of logs gets more complicated as the lithology becomes more complex. The type of hydrocarbons expected may also be taken into account.

As discussed previously, the Sonic, Density and Neutron Logs may be affected by mineral mixtures in the formations and gas. Therefore, when the formations drilled vary from sandstones to limestones to dolomites and varying mixtures of each, it becomes important to use a combination of logs that can help detect the variations. It also becomes important to detect the presence or absence of gas.

The effect of gas on the Sonic and Density Logs is to make the porosity appear higher than it truly is. The gas effect on the Neutron Log is to make the porosity appear lower than it truly is. Therefore, the combination of Neutron — Sonic or Neutron — Density is very helpful to detect the presence of gas. The one most generally selected is the Neutron-Density.

A technique called cross plotting (explained later) can be used on the Neutron-Density combination to detect variations of lithology so that proper analysis of the porosity may be made.

In the case of carbonate reservoirs there may be secondary porosity involved also — such as fractures, vugs, solution cavities, etc. If so, the Sonic Log used in conjunction with the Neutron-Density combination can be helpful.

The first example of log analysis will involve an Induction — Electrical Log and a Sonic Log run by Schlumberger, through a sandstone interval. The logs are shown in Figures 51 and 52. The interval of interest is from 3,773 feet to 3,795 feet. The Induction — Electric Log is Figure 51. The spontaneous potential is the curve in the left track, with the shale line about two divisions left of the depth column. The deflection in the interval of interest is to the left from the shale line. This is negative S.P. and indicates

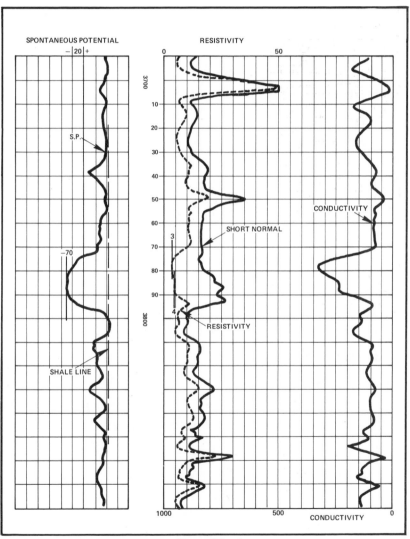

Figure 51

105

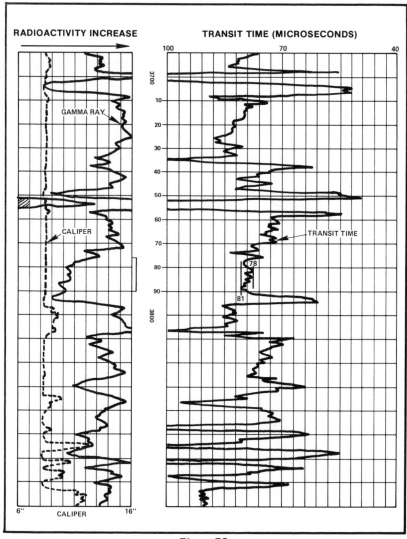

Figure 52

106

that the mud filtrate is less saline than the formation water (high Rmf/Rw). The curves in the center track of the log are resistivity measurements. The solid curve is the 16-inch normal, a shallow investigating device designed to measure predominantly in the invaded zone. The dashed line is the deep investigating curve designed to measure beyond the invaded zone to determine true resistivity. This curve is derived by using the curve in the right track of the log. The instrument used to obtain this log is designed to measure formation conductivity rather than resistivity. The conductivity measurement is recorded in the right track and may extend into the center track. This value is converted to resistivity by the surface control equipment using the relationship $R = 1/C$ (resistivity is equal to 1 divided by conductivity).

The curves on this log are scaled as follows: since the spontaneous potential is a measurement of the very small voltage variations in the borehole, it is scaled in millivolts, with each division in the track equal to 20. The maximum deflection in the sand interval is about three and one-half divisions left of the shale line so the S.P. is equal to -70 millivolts. The resistivity scale in the center track is zero at the depth column and increases by 5 ohm-meters per division to the right side of the center track (there are 10 divisions so the scale is 0-50).

The unit of conductivity is the opposite of ohm (the unit of resistivity) or mho. Formation conductivity covers an extremely broad range, so to facilitate the scaling the values recorded are in millimhos (one-thousandth of a mho), with zero at the right side of the long and 1,000 millimhos at the depth column. This allows the measurement of 1 ohm-meter to infinity without the curve going off the log. Also, with zero at the right side the resistivity curves and the conductivity curve deflect in the same direction for variations in formation resistivity and conductivity.

Figure 52 is the Sonic Log with a Gamma Ray and caliper. The caliper is the dashed line in the left track and indicates hole size changes from 6 inches at the left side of the track to 16 inches at the depth column. The Gamma Ray is the solid line in the left track. Radioactivity increases from left to right, with the shales having the highest radioactivity. The interval of interest has radioactivity considerably less than the shales, but with a little shaliness indicated in the upper portion. The Sonic Log is recorded in the two right tracks with the lowest transit time being at the right side of the log. Transit time is measured in microseconds (millionth of a second). Most formations will transmit sound between 40 and 100 microseconds-per-foot, so the log is scaled with 40 microseconds at the right side of the log, 70 microseconds at the center of the two tracks, and 100 microseconds at the depth column. Each log division equals three microseconds.

The first step in analyzing these logs will be to solve the Sonic Log for porosity. To do this the following formula is used:

$$\emptyset = \frac{\text{Delta-t log} - \text{Delta-t matrix}}{\text{Delta-t fluid} - \text{Delta-t matrix}}$$

where Delta-t log is the reading from the log in microseconds per foot, Delta-t matrix = transit time of the matrix rock at zero porosity, and Delta-t fluid = transit time of the fluid in the pore spaces (usually water).

Velocities and transit times for various reservoir rocks are: sandstones, 18,000 to 19,500 feet-per-second with transit times equal to 55.5 to 51 microseconds; limestones, 21,000 to 23,000 feet-per-second with transit times equal to 47.5 to 43.5 microseconds; dolomites, 23,000 to 26,000 feet-per-second with transit times equal to 43.5 to 38.5 feet-per-second. Water has a velocity about 5,300 feet-per-second with transit time equal to 189 microseconds.

One can use longhand, a calculator or a slide-rule to solve the formulas used in log analysis, but for convenience the service companies provide chart books with charts that do the mathematics for you. The Schlumberger chart to solve the Sonic Log is shown in Figure 53.

In Figure 52 the transit time in the zone of interest is from 78 to 81 microseconds, with an average of about 80. Entering these values on the scale at the bottom of Figure 53 and projecting them upward to the 18,000 feet-per-second line (for sandstones) the left to porosity scale shows porosity from 17 to 19 percent with the average being about 18 percent. This chart has 189 microseconds for fluid transit time built into it. Fourteen

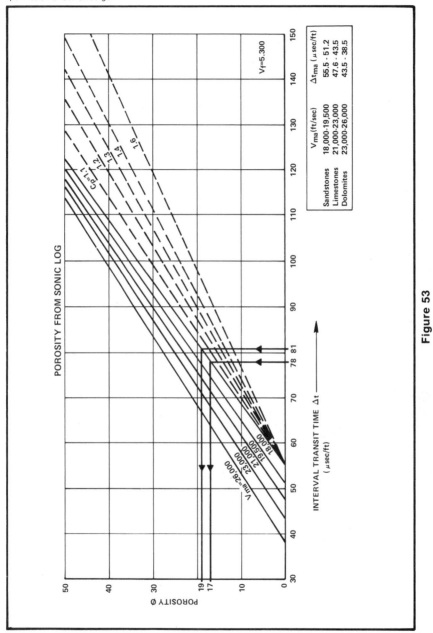

Figure 53

feet (3,776 feet to 3,790 feet) is sufficient to make the zone interesting enough to pursue further.

The formula for determination of water saturation is:

$$Sw = \text{square root of:} \sqrt{\frac{F \times Rw}{Rt}}$$

F equals the formation factor derived for porosity, Rw equals the resistivity of the formation water, and Rt equals the resistivity of the uninvaded part of the reservoir. Figure 54 will solve this relationship for us. As mentioned previously there are many catalogues of formation water resistivities, and Rw for this formation is equal to .035 at 115° (temperature at the depth of the formation). From the dashed curve on Figure 51, Rt can be read as 3 to 4 ohm-meters. F can be determined from Figure 54-stem 5 or Figure 57, and for 18 percent, porosity, F equals 25.

On Figure 54, find .035 on stem 4 (Rw), and F equals 25 (∅ = 18 percent) on stem 5. With a straight edge, align these two points and draw a line from stem 4 through stem 5 to stem 6. This establishes the value .87 which is the resistivity of the zone if Sw were 100 percent (Ro). Find the points 3 and 4 for Rt on stem 7. Align the straight edge from .87 on stem 6 with 3 on stem 7 and draw a line through them to stem 8, then from .87 through 4 also. The two numbers on stem 8 are 47 to 54 percent. This is the water saturation range for this sandstone, which indicates the presence of hydrocarbons. This zone was completed with initial production being oil and water.

The second example is a bit more complicated in that it involves a Dual Induction — Guard Log and a Compensated Density-Neutron Log run by Welex. The Dual Induction — Guard Log by Welex and the Dual Induction — Laterolog by Schlumberger are similar type logs. The logs are shown in Figures 55a and 55b. The interval of interest is a sandstone from 10,580 feet to 10,656 feet. The Dual Induction — Guard Log is in Figure 55a. The Spontaneous Potential curve is, as usual, in the left track, with the shale line about two and one-half divisions left of the depth column. The deflection in the interval of interest is to the left of the shale line about three divisions (45 millivolts). This is negative S.P. and indicates that the mud filtrate is less

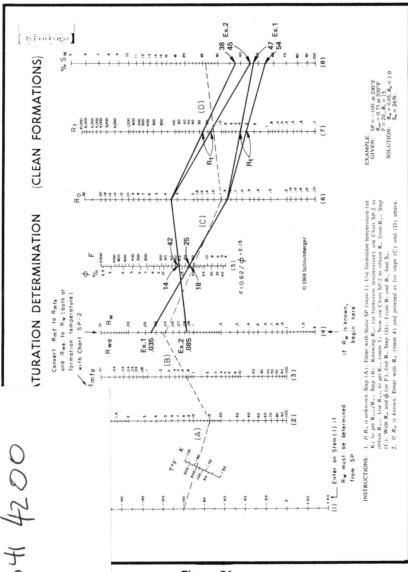

Figure 54

111

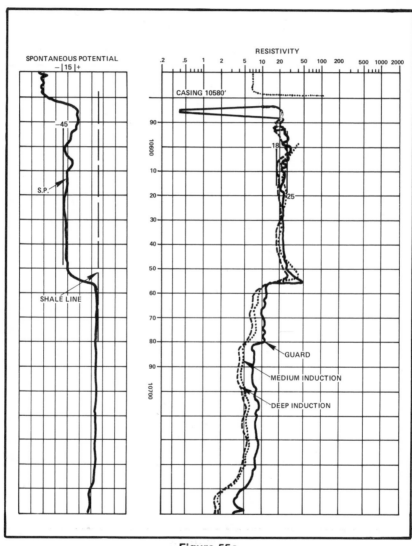

Figure 55a

112

saline than the formation water (high Rmf / Rw). The curves in the right track are resistivity measurements. Please note that the scales for the curves in the right track are logarithmic rather than linear; this is beneficial in some analysis techniques. It also permits a very broad range of resistivity measurements and still keep the curves within the two tracks. The solid line is the guard log which is designed to measure in the invaded zone. The small dashed line is the medium induction log which is designed to measure deeper into the formation, but may still be affected by the invasion of filtrate. The heavy dashed line is the deep induction log which is designed to measure beyond the invaded zone and help determine true formation resistivity (Rt).

In Figure 55b the line in the left track is the Gamma Ray Log, and the curves in the right tracks are the Compensated Density and Neutron Logs. The solid line is the neutron and the dashed line is the density. In the zone of interest the Density Log indicates porosity considerably higher than that indicated by the Neutron Log. This could cause concern as to what the porosity actually is. A technique called cross plotting is used on these logs in order to arrive at the correct solution.

As mentioned previously, the Density and Neutron Logs are affected by the mineral mixtures of the formations, therefore some knowledge of the mixture is necessary in order to solve the logs accurately for porosity.

The Neutron Log equipment has been calibrated in the laboratory to respond to limestone in a linear fashion through the normally expected porosity range. The control equipment may then be adjusted to record porosity for sandstones or dolomites.

The Density Log equipment has been calibrated to respond linearly to formation densities between 2.0 and 3.0 grams-per-cubic-centimeter. The control equipment may be adjusted to record porosity for formations using the following relationship:

$$\text{Porosity} = \frac{\text{P-grain} - \text{P-log}}{\text{P-grain} - \text{P-fluid}}$$

P-grain equals the density of the formation at zero porosity; P-log

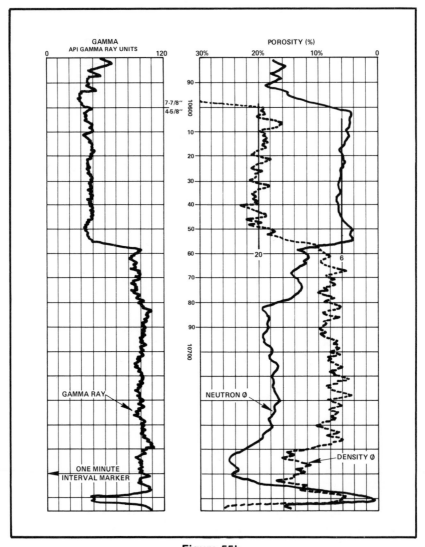

Figure 55b

114

equals the density read from the log; and P-fluid equals the density of the fluid in the pore spaces. Grain densities for reservoir rocks are: sandstones equal 2.65 to 2.68, limestones equal 2.71; and dolomites equal 2.87 (all are grams-per-cubic-centimeter).

When these logs are run in combination, the surface control equipment is adjusted to record them both as limestone porosity. If the two logs indicate the same porosity value, it means the rock being measured is limestone. If the Density Log indicates porosity less than the Neutron Log, it means the rock being measured is heavier than limestone (probably dolomite). If the Density Log indicates porosity greater than the Neutron Log, it means the rock being measured is lighter than limestone (probably sandstone). Of course there can be many combinations of these mixtures, therefore cross plotting can be extremely valuable to determine when these mixtures occur. The cross plotting technique utilizes a chart to solve the two logs. Such a chart is Figure 56.

Also as previously stated, gas can cause an effect on the response of these logs. The effect of gas on the Density Log is to cause it to give an apparently too high porosity. The effect of gas on the Neutron Log is to cause it to give an apparently too low porosity. The ultimate effect on the cross plotting technique is to cause the points being plotted on the chart to be displaced upward and left of where they should fall. The direction of correction is shown by the arrow in the upper left part of the chart. The amount of correction depends on the lithology involved.

Let us now proceed with the analysis of these logs. The first step will be to determine whether or not there is sufficient porosity to warrant further work. In order to demonstrate the technique and procedure we will use average values for porosity from each log. In actual practice many points will be chosen and cross plotted, especially in an interval as thick as this one. This permits maximum and minimum values, as well as a predominance of values to be determined.

In Figure 55b average values of porosity for each log may be picked as follows: the Density Log equals 20 percent, and the Neutron Log equals 6 percent. Turning to Figure 56 these values can be found on the appropriate

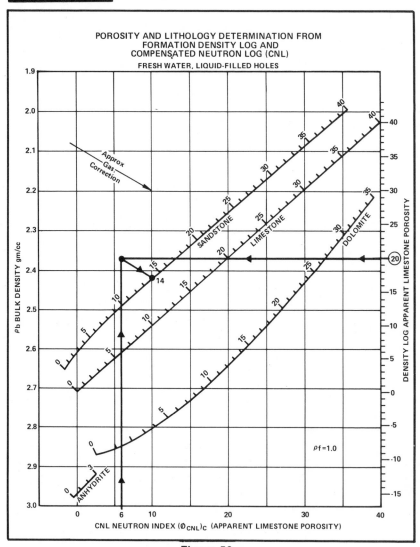

Figure 56

scales. The Neutron porosity at the bottom, and the Density porosity at the right side. When the points are found, project them into the chart until they intercept. In this case this point falls to the left of the three lithology lines. This immediately indicates there is a gas effect on these logs. The direction of correction is indicated by the arrow on the chart. How far to be corrected is determined by the lithology. This zone happens to be a sandstone, so from the point found move parallel to the gas correction arrow to the sandstone line. Correct porosity is then found where the new point falls on this line. It is equal to 14 percent.

The sudden shift on the logs at 10,600 feet is due to the hole size change from seven and seven-eighths to four and five-eighths inches. This interval is 76 feet thick so it definitely is an interesting zone. The next step is to determine the water saturation (Sw).

To do that we must determine true resistivity (Rt) from the Dual Induction — Guard Log (Figure 55a). Please note that all three resistivity curves are closely grouped together. This generally occurs when the formation has very little invasion by the mud filtrate, so that all curves tend to read deeper than the invaded zone. In a case like this, Rt can confidently be determined from the deep induction curve. The deep induction curve (heavy dash) varies around 20 (from about 18 to 25).

The next item necessary before entering the Sw formula is Rw. From a catalogue of water resistivities we can find a value for this zone in this area equal to .085 at 175° (temperature at formation depth).

We now have Rw = .085 @ 175°, Rt = 1 to 25, and $\emptyset$ = 14 percent, for which we need formation factor (F). On Figure 54 or Figure 57 we find F = 42.

Either solve the Sw formula as presented earlier or enter Figure 54-stem 5, which will do it for us. On Figure 54 find .085 on stem 4 and 42 ($\emptyset$ = 14 percent) on stem 5. Align these two points with a straight edge and draw a line from stem 4 through stem 5 to stem 6, which gives 3.5. This is the resistivity of the formation at 100 percent Sw (Ro). Find 18 and 25 on stem 7 and align the straight edge with 3.5 on stem 6 and 18 on stem 7 and draw a line to stem 8, repeat through 25 on stem 7. Find the values 38 and 45 on

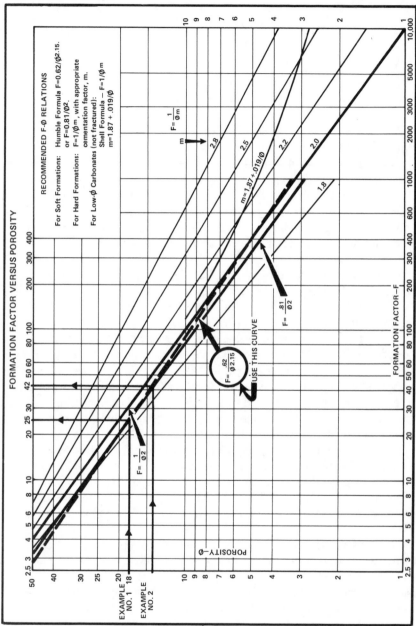

Figure 57

stem 8. This is the Sw range of this zone, and indicates the presence of hydrocarbons. The zone was completed as a gas well with condensate.

These two examples are relatively straight-forward cases of log analysis. There can be and generally are much more complex situations which tax the knowledge and expertise of the log analyst. He must use all the help he can get from the geologist, petroleum engineer, and production people pertaining to the rock characteristics, fluid mechanics, and completion problems when making decisions about producibility of a formation.

Of course this is not the only use for well logs. They are used extensively by geologists as a source of information for mapping. They correlate various formations versus depth, determine interval thickness between zones, and get zone thickness variations for example. This helps them make structure maps, cross sections, isopach maps, etc.

Petroleum engineers will use logs to compare formation characteristics between wells as a field is developed so that they may exploit the reservoirs more efficiently.

Production people use logs to help them in completing the well, and to help solve problems should they arise.

Drilling contractors sometimes use logs to help them plan drilling programs, what type bits to use, when they may encounter shale problems, proximity to high pressure zones, etc.

Well logging and the analysis thereof has grown to such an extent that it is now one of the very important parts of the petroleum industry.

It is hoped that this chapter has been helpful in promoting some understanding of this fascinating field.

Chapter Five

MAPPING METHODS

Section One: Surface

§ 5.01 Areal Distribution.[1]

Strictly speaking, a geologic map is a surface map of the distribution of various formations in a given area. In colored or other symbolic patterns, the bands of different formations are shown where they outcrop at the surface. Strike and dip symbols are used to indicate the direction and amount of dip of the beds. The symbols are placed at the location of the measurement which was made on the ground. Figure 58 is an example of a portion of a geologic map.

The strike of a bed is the compass direction which a dipping bed takes as it intersects a horizontal surface (or any mapped horizon). The dip is the number of degrees from horizontal that the formation deviates. In Figure 58 the arrow points to a place where the strike is roughly northeast-southwest and the dip is 12 degrees in a southeasterly direction. Just to the east of that

---

[1] "Areal" should not be confused with "aerial." The former refers to area and the latter refers to the air, as in aerial photographs.

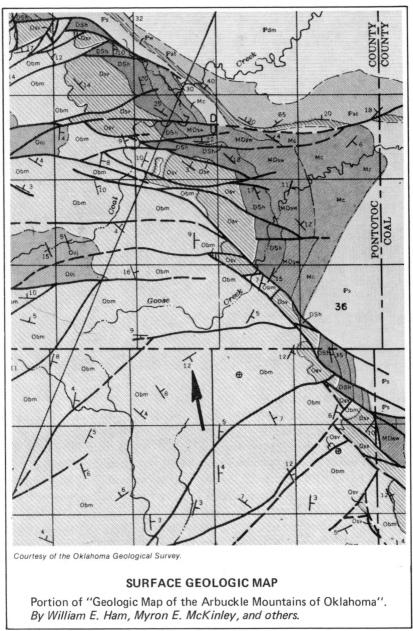

Courtesy of the Oklahoma Geological Survey.

SURFACE GEOLOGIC MAP

Portion of "Geologic Map of the Arbuckle Mountains of Oklahoma".
By William E. Ham, Myron E. McKinley, and others.

Figure 58

122

symbol there is another symbol indicating that there is no dip — the beds are perfectly flat at that locality.

The formations are labeled with abbreviations of their names such as Obm in the illustration which stands for Ordovician Bromide-McLish Formations. Faults are shown but there are no contours.

The formation boundaries or contacts are shown on all geologic maps. Sometimes the boundary lines are dashed if there is some uncertainty. This could happen if some river alluvium covered up the contact in a few places. The standard geologic map is prepared by going over the ground, on foot, taking strike and dip measurements, and locating contacts and faults on a base map. Aerial photographs are often used as an aid in mapping. Sometimes the mapping begins with the photos and concludes by field-checking certain parts of the area to confirm the photo-interpretation.

§ 5.02 Topographic Maps

Another surface map often used by geologists is the topographic map, which is a contour map. The contours are drawn to represent the varying elevations of the earth's surface regardless of the formations involved. These too are prepared with the aid of aerial photographs. The U.S. government publishes these maps at various scales and they are available for most areas, but not all.

There are a number of uses for topographic maps. Because they show the terrain in some detail, they are excellent for picking drill-sites. Drainage is also shown in great detail as well as man-made features such as roads, dams, canals and mines.

Although geologic formations are not mapped, there is a lot of geology to be seen on topographic maps. People often laugh at "creekology," but there can be no doubt that the courses of many streams are controlled by fractures, faults, formation changes and subsurface structures. Radial drainage, for instance, forming a sort of circle in an area which otherwise has, say, dendritic drainage, is an anomaly and may reflect a subsurface dome. Another example of a drainage anomaly would be an unusually straight portion in a river channel which usually meanders. The straight

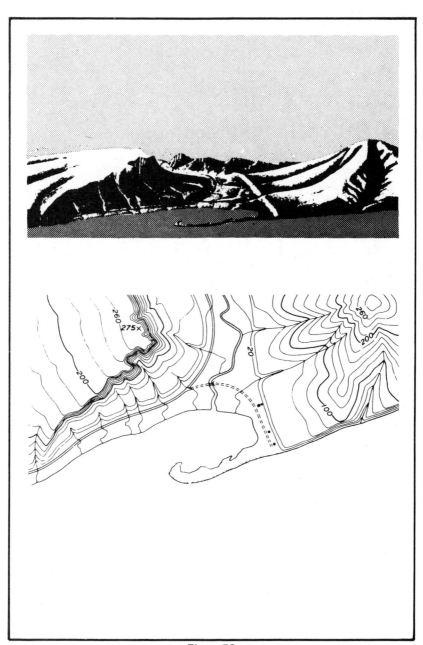

Figure 59

portion may be following along a major fracture. Geologists use these anomalies as clues to aid them in a more complete interpretation of a region or locality.

Of course, there are many non-geological uses for these maps, such as for hunting, fishing, camping and hiking. They are also used by city planners, soil conservationists, foresters, flood control engineers, and many others. An example of a topographic map is shown in the lower part of Figure 59. the upper part of the illustration shows a bird's eye view of the same area. The clearest explanation of the way a topographic map is made is given by the U.S. Geological Survey as follows:

"Some of the features depicted on a topographic map are illustrated in the accompanying birds' eye view of a river valley and the adjoining hills. The river flows into a bay which is partly enclosed by a hooked sandspit. On both sides of the valley are terraces through which streams have cut gullies. The hill on the right has a smoothly eroded form and gradual slopes above a wave-cut cliff. whereas the one on the left rises to a steep slope from which it falls off gently and forms an included tableland crossed by a few shallow gullies. An unimproved dirt road and bridge provide access to a church and two houses situated across the river from an improved light-duty road which follows the sea coast and curves up the river valley.

The lower illustration shows the same features represented by symbols on a topographic map. Elevations are represented by contour lines; the vertical difference between contours in this illustration is 20 feet.

To understand the contour symbol, think of it as an imaginary line on the ground which takes any shape necessary to maintain a constant elevation above sea level. The shoreline shown on the map illustration is, in effect, a contour representing zero elevation or sea level. If the sea should rise and cover the land, the shoreline would trace out, in turn, each of the contour lines shown on the map. Since the vertical difference in elevation between contours in this example is 20 feet, the shoreline would coincide with a new contour each time sea level rose 20 feet.

For easier reading, index contours (every fourth or fifth contour, depending on the contour interval) are accentuated by making the lines heavier. Supplementary contours, used to depict features which the basic contours do not adequately portray, are shown as dashed or dotted lines. Figures in brown at intervals along contour lines give the elevations of the lines above sea level. The elevation of any point can be read directly, or interpolated between contours. Map users who are concerned with quantitative measurement of terrain features can determine this basic data from map contours."

§ 5.03 Remote Sensing

The space age has given us the technology to do a great many new things to study the earth. Not only photographs from space, but also imagery such as radar and infra-red, now are available to the general public and geologists have had a field day learning to use these new tools.

Because of the distance involved, large areas of the earth's surface can be seen on one moderate-sized piece of paper. Although the remote-sensing images are essentially a reconnaissance tool, they have shown a surprising assortment of new geological features which were not previously recognized. Lineaments of great length are an example; they probably are indicators of deep-seated faulting.

There are tonal anomalies, too, caused by some alteration of the surface — perhaps by leaching of minerals or oxidation. All of these anomalies have some geological explanation and they are often clues to exploration for oil and gas, as well as other minerals. The following are some of the mapping devices in use today:

Side-Scan Sonar. The principal geological use for side-scan sonar is to map the sea floor. Besides topography, the system can depict geology if rocks are exposed rather than being covered by recent sediment. Gas seeps can also be identified.

Landsat Data. Formerly called ERTS Imagery, this information is produced from our earth-orbiting satellites. The equipment is basically a multi-spectral scanner which gathers data in several different wave length regions of electromagnetic energy, including infra-red.

This information is useful in a large number of fields, such as forestry and agriculture. In geology, it has shown up many lineaments and other structural features which had not previously been recognized. Other new features are areas of mineralization and lithologic mapping.

Side-Looking Radar. This technique is usually called SLAR, for side-looking airborne radar. It functions in the microwave region of the electromagnetic spectrum. The imagery obtained looks much like aerial photography. The advantages of SLAR are that the surveys can be flown at night or on cloudy days without adverse effect on the imagery.

Aerial Photography. With our newer, more sophisticated sensors in wide use, there is a tendency to forget that aerial photographs, too, are remote sensors. They are widely available, relatively inexpensive, and extremely useful. The most common are in black and white, but color is available. Among the features which give clues to the geology are tonal changes, in vegetation or rocks, lineaments, drainage patterns, and vegetation distribution patterns.

To summarize geological mapping of the surface of the earth, especially in the United States, one might say that it has come full circle. From the early days of the oil business when surface maps were about all we had, to a middle period when subsurface maps were the only things, we have come back to another and more informative look at the surface.

Section Two: Subsurface

§ 5.04 Areal Distribution

Just as maps can be made of the surface showing the areal distribution of formations, similar maps can be made of the subsurface.[1] These are usually made at unconformity surfaces and almost always are based solely on well-control points. Seismic information is not very well adapted to mapping the distribution patterns but it can be of significant help in making a contour map of the unconformity surface.

A subsurface areal distribution map is shown usually in color patterns to represent the different formations. Sometimes contours are superimposed on the color pattern. These contours, strictly speaking, represent topography on the old erosional surface. However, there is usually an element of the structural in such a contour map because any structural event occurring after the unconformity would fold and deform the unconformity surface. That often is one of the problems in dealing with an unconformity surface: How much erosional topography are we seeing and how much post-unconformity structure?

§ 5.05 Structure Maps

The commonest subsurface map in the oil business is the structure map. It is constructed in the following manner:

Logs of wells in the area of interest are gathered and correlated. Accurate correlation is crucially important. A zone which occurs in all the wells is picked to be the mapped unit or formation. This can be anything — a shale bed, a sandstone or a limestone. Most geologists like to use limestones, as they are frequently more persistent than sandstones. The zone to be mapped, if it is not the objective pay itself, should be stratigraphically close to the object pay — for example, within a couple of hundred feet. It may be below or above the pay, depending on the local geology.

The formation top is picked. From the depth, read off the log, the elevation must be subtracted unless the well is at sea level. The resultant number is posted on a map beside the well symbol. When all available data are posted, the contours are drawn and the structure map is finished. The actual drawing of the contours can be easy or difficult, depending on the number of control points and their distribution.

There may be several ways to draw the contours rather than just one.

In that case, the geologist must use his or her best judgment as to the most appropriate way to draw them. In the interpretation of oil and gas prospects, it is perfectly appropriate to use imagination provided it is based on sound geology. This is true for all contoured subsurface maps, not just structure maps.

Structure maps are usually drawn on a regional basis first, then a local basis. These maps may be based solely on subsurface data, or may have seismic information included.

Seismic structure maps are often prepared separately, partly because the increased amount of data available requires more map space. A popular scale for seismic maps is 1 inch equals 2,000 feet. This allows the information to be posted for the four to six shot-points per line-mile, which is the usual spacing.

§ 5.06 Isopach Maps

The isopach map depicts the thickness of some bed or formation, such as a sand bar. The contour lines are drawn through points of equal thickness. The term isochore is used if the interval mapped encompasses more than one unit, but in practice the term isopach has been used in place of isochore as a universal term for thickness map.

A major use of isopach maps is to delineate features, such as sand bars or river channels, which have lateral and vertical limits. The goal is to define the shape, or geometry, of the sand body in order to predict the thick part of the channel or bar, so that better reservoir conditions can be expected.

Another use for thickness maps is a more regional one, wherein one surface involved is an unconformity surface, usually the top of the mapped interval. The erosion which creates the unconformity is differential erosion, meaning that high areas, or hills, get eroded more rapidly than low areas. Using this premise, old structures can be assumed to suffer more erosion than flat-lying or low areas. The unconformity surface is created, followed by more deposition. Later mapping of some thickness related to the unconformity may reveal a buried structure which had not been previously recognized.

§ 5.07 Initial Potential Map

In the 1940's and 50's it was a common practice in some areas to drill just a few feet into the producing sand and complete the well without ever drilling all the way through the sand. Thus, we rarely knew the geometry of the sand body and had difficulty in predicting the sand distribution pattern.

In an effort to overcome this problem, the Initial Potential Map was devised. It was based on the assumption that completion practices were uniform, and that reportage of initial potentials was accurate and comparable. The resultant map can be an indicator of the geometry of the producing sand, because the higher IP's usually will occur in the best portion of the channel or bar.

This is not recommended as a useful map for modern wells, or at least such a map should be taken with a grain of salt. But for old areas, in an effort to map the thickness of sand bodies which were never completely penetrated, it can be a useful technique.

§ 5.08 Ultimate Production Maps

Like the Initial Potential Maps, the maps which show ultimate recoverable production are an attempt to portray something about the reservoir rock without using direct measurements of the rock. As always, the map is for the purpose of prediction.

The mapping of ultimate production means the mapping of estimated figures and, therefore, the result can be very "iffy." However, such a map can be of some help and should be considered as supplemental information. In limestone reservoirs, for instance, the porosity and permeability and overall characteristics of the reservoir may seem quite uniform. Yet some producers are substantially better than others. This may be due to an unrecognized fracture pattern which may be at least hinted at in an IP Map or an Ultimate Production Map. Used with care, the map can point to trends for exploration purposes and can also be used for the planning of development locations.

§ 5.09 Facies Distribution Maps

There are several forms that facies maps can take. One of the simplest has no contours but simply shows where a given formation is limestone ("limy") and where it is sandstone and where it is shale. The map is prepared by using either colors or symbolic patterns for the different kinds of rocks. This is a particularly useful map to work up in the pursuit of stratigraphic traps because the boundary, for instance, between a limestone and a dolomite facies can be one aspect of a trapping mechanism.

Another type of facies map shows the ratios between sand and shale (clastic ratio) or limestone and dolomite (carbonate ratio). Again the purpose is to predict changes in lithology which may have an effect on the trap.

Chapter Six

CROSS SECTIONS

A cross section is a profile or slice down through the earth which shows two dimensions, one vertical and one horizontal. Using a cross section together with a map, one can put together a three-dimensional picture of a prospect or an area. There are three variables in the construction of a cross section which must be considered: there are the choices between true scale vs. exaggerated scale; structural vs. stratigraphic presentation; and electric log vs. stick section. Each of these is described below.

§ 6.01 True Scale

A true scale cross section is one in which the vertical scale is the same as the horizontal scale. For instance, if one inch of paper equals 500 feet vertically, then one inch of paper will equal 500 feet horizontally also. The important advantage of using true scale is that when one looks at the cross section the dips of the formations are approximately correct as shown. This type of cross section is used most often in areas of deep drilling and steep dip, and rather than electric logs, "sticks" symbolic of the wells are used. Another advantage of the true scale section is that one can take measurements directly from the cross section very easily. For instance, dip can be measured directly with a protractor. And, intervening distances can be measured with a ruler without having to convert any numbers due to exaggeration. Figure 60 illustrates the true scale section using sticks instead of logs.

§ 6.02 Exaggerated Scale

The use of exaggerated scale is usually just a practical matter — it condenses a lot of information onto a manageable piece of paper. It is most commonly used when electric logs are used in the cross section. The standard small scale log has a scale of 1 inch = 100 feet. That is, one inch of paper (vertically) represents 100 feet of drilling. Such logs are often reduced down further so that they have a scale of, for example, 1 inch = 200 feet. But, when these logs are used in a cross section, there may be some miles between wells and, therefore, a true scale of 1 inch = 200 feet horizontally isn't feasible. So, some multiple of the vertical scale is used. Often it is 10 times the vertical, or 20 times, and is usually shown in the legend as:

10X Exaggeration

If the exaggeration is 10X and the vertical scale is 1 inch = 200 feet, then the horizontal scale is 1 inch = 2,000 feet.

This type of presentation is also used in areas of low dip because the feature to be shown, such as an updip pinchout, might be almost invisible as true scale.

§ 6.03 Structural Section

A structural cross section is one in which the formations are shown in their true position relative to sea level. The datum plane or reference plane is sea level and additional reference lines are usually drawn to show where the formations fall, either above or below sea level. Structural cross sections are used to show structural relationships, although stratigraphic relationships also may be included. Figures 39 and 60 are structural cross sections.

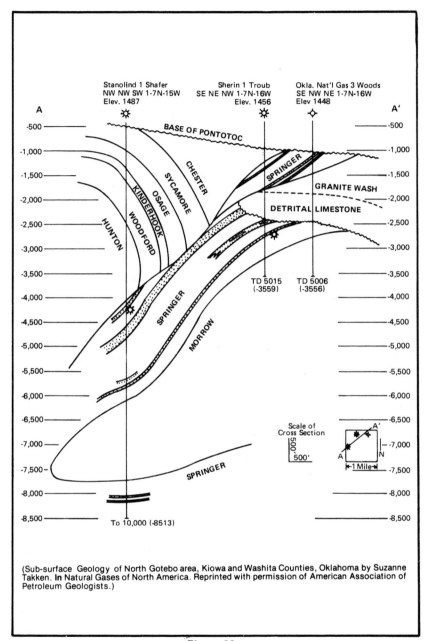

(Sub-surface Geology of North Gotebo area, Kiowa and Washita Counties, Oklahoma by Suzanne Takken. In Natural Gases of North America. Reprinted with permission of American Association of Petroleum Geologists.)

Figure 60

135

§ 6.04 Stratigraphic Section

The purpose of this presentation is to show stratigraphic relationships only. Little or no structural information is conveyed. A datum is selected which is a formation common to all wells in the prospect area and usually it is within a few hundred feet (drilling depth) of the specific horizon to be shown. Often a pinchout or other facies change is to be demonstrated. It is common to use electric logs for this. The datum plane in this case has nothing to do with sea level, but is the zone to which all wells are related. We say the cross section is "hung" on the Topeka Limestone, for instance.

§ 6.05 Electric Log Sections

The great majority of cross sections these days are constructed using electric logs. This is because they are almost universally available and because the curves can be seen which were used in the correlations. Therefore, one can look at the cross section and determine for oneself if one agrees with the correlations. That is the great advantage of the electric log cross section. The disadvantage is the size of the log, which virtually insures that there will be some exaggeration in the cross section. Figure 39 is an example in which, in order to achieve true scale, the logs had to be reduced almost to illegibility.

§ 6.06 "Stick" Sections

The term "stick" refers to the fact that a single line or "stick" is used to represent the well, rather than the actual electric log. The advantage and disadvantage are just the reverse of the electric log section. One cannot see the curves of the logs so cannot verify at a glance whether one agrees with the correlations. On the other hand, a stick section allows for true scale and can be drawn on a very small sheet of paper which is convenient for reports. Figures 37 and 60 are examples.

Most cross sections are drawn along virtually straight lines, such as east to west, or north to south, or some direction in between. The wells selected for the cross section should fall into a relatively straight line. If one well is quite far off the line, it is often suitable to "project" that well into the line of cross section. It is preferable to do this than to zig-zag the line of cross section all over the map, because the observer in the latter case has a hard time assimilating the information properly.

One exception to that, however, is the fence diagram. That is a form of cross section which actually is many cross sections which often intersect one another and in this presentation look like so many fences all over the map. Here again, the intent is to convey three dimensions. These are informative illustrations but are rather tedious to prepare.

A related illustration is the block diagram which is an even more elaborate attempt to show three dimensions. It takes an artist's feel for perspective as well as the ability to visualize three dimensions easily. Not many petroleum geologists take the time to construct block diagrams these days.

In summary, cross sections are constructed to show certain geological information, and there are various ways to do this depending on what the information is. Cross sections should be an integral part of every prospect presentation. Quite often, a cross section gives a clearer picture than a map view.

Chapter Seven

WHAT TO LOOK FOR
IN A PROSPECT

§ 7.01 Examining Submittals

As a general rule, a prospect should have a geological and/or geophysical presentation in support of the prospect, a prospect outline showing the land position, an estimate of the risks, reserves and general economics, and a statement regarding the terms of the deal. Each of these aspects is discussed briefly below.

§ 7.02 Geology and Geophysics

It is customary, but not essential, to provide a regional map so that the general setting of the prospect can be seen. If all people involved are thoroughly familiar with the area, then a regional map isn't necessary.

Bear in mind what kind of prospect you are being shown. If it is a structural prospect then some kind of structure map should be presented. A subsurface geological map is usually a first approximation. A structure map of the same area based on seismic data is very desirable — some would say essential. The amount of closure shown on the prospect is the important part of the structure map. (Structural closure vs. productive closure is discussed in detail in Chapter Two.) Suffice it to say here that there must be closing contours if the prospect is truly a structural trap. The lowermost closing contour usually serves as the prospect outline although the "buy" outline may be somewhat larger.

Frequently in older producing areas, where there is a lot of history to look at, there develops a rule-of-thumb about closure. Such "rules" are like the following: "We need at least 150 feet of closure in this area" (or it could be 50 feet or some other number). These rules-of-thumb are very useful, but not foolproof. Many a field has produced with either more or less closure than predicted. Also, the geophysical interpretation need not be a carbon copy of the geological interpretation, nor vice versa. They should be close, however. Each supports the other.

In addition to a structure map, there should be a cross section across the prospect showing the proposed location in true scale if possible. If the cross section is a "stick" section, then a type electrical log should be included in the prospect materials.

If there is more than one objective, the maps and cross sections should show each one. In summary, for the structural trap we can list a number of exhibits which could, and perhaps should, be included:

1. Regional map
2. Local structure map
3. Structural cross section
4. Type log
5. Prospect outline and land position

With stratigraphic prospects, similar exhibits are required, but a structure map, usually submitted by tradition, is not very informative and need not be included. An electric log cross section and an isopach map are the most common geological exhibits for stratigraphic traps. Seismic work may or may not be helpful in defining strat traps.

Look for the control points on the maps to see how much information is available. Is the nearest well two miles, or 2,000 feet, away from the prospect? Is the correlation obvious and easily determined, or is it nebulous and doubtful? These are the factors which go into the evaluation of risk.

§ 7.03 Risks

There are two aspects of geological risk which are important to consider. These risks are not those of a statistical nature, such as that only one in 12 wildcat wells actually finds oil. The risks which first must be taken into consideration are: (1) Will the objective formation actually be present? (2) Will there be an oil and gas trap?

Item number one, which involves the presence or absence of the objective formation, is truly a geological risk, usually related to stratigraphic traps. Consider the Red Fork sand (Pennsylvanian) of Oklahoma or the Muddy sand (Cretaceous) of Wyoming. These are stream deposits, meaning that their distribution is quite erratic and, therefore, there is a possibility of not getting any sand at all when drilling a wildcat well. On the other hand, there are many objectives which are so uniformly deposited that there is no doubt at all that the drill will encounter the formation. Examples would be the Ellenburger of Texas, the Hunton of Oklahoma, and the Phosphoria of Wyoming.

The second risk is the most basic one of all: is there a sufficient trap and will oil and gas actually be trapped there? It is quite possible to find a situation where oil and gas has been trapped at one time and later the trap leaked or was somehow flushed — possibly by changing hydrodynamic pressures. Evidence for the fact that hydrocarbons were once there is the presence of residual oil in cores or cuttings.

In any given prospect one or both of the above-mentioned risks will be present. There is no easy way to put a value on these risks although some people occasionally try. An experienced geologist can provide an educated guess as to the odds of finding the sand or finding the trap, but it will still be a guess. Formal risk analysis using probability theory is indeed used in an effort to quantify the risks encountered in exploration, but it is undoubtedly best applied to multiple-well programs.

§ 7.04 Reserves

Estimating reserves in advance is, of course, a difficult business, but a necessary one. In a producing area which has some history, the best method is to tabulate the cumulative production from existing similar fields, make an estimate of the ultimate primary recovery, take an average and apply it to the new prospect.

The volumetric method uses estimates of porosity, drainage area, and recovery factors which seem appropriate for the area. After a successful wildcat is drilled, actual figures for some of these parameters can be used. There are still some uncertainties, however. The principal one involves the area which one well will drain. In the early days of the oil business the wells were drilled unnecessarily close together, which was a wasteful practice. Now, with formal spacing rules in effect, many wells are drilled a mile or more apart and it could be that one well may not be able to drain 640 acres. The calculations for the volumetric method of estimating reserves per acre-foot are given in Appendix 3.

Gas reserves are particularly difficult to estimate because the physical behavior of gas is quite variable. Pressures, temperatures, and amount of liquids in the reservoir all play a role and yet are difficult to predict in advance. After the well has produced for a while, decline curves can be used. These are charts with a plot of cumulative production versus pressure. When several points have been plotted a curve can be drawn and then extrapolated to some arbitrary abandonment pressure. As usual, this method is not as simple as we would like it to be. A compressibility factor should be used, for instance, in the pressure plot. An unexpected influx of water in the reservoir, a change in pipeline pressure and other factors, can alter the amount of recoverable gas which will ultimately be produced.

Sometimes rules-of-thumb are as useful as careful calculations since there are so many estimates and unknowns in the latter. Here is one which is applicable in the Midcontinent:

For each mile of depth, multiply the porosity of the reservoir by 50 to get an estimate of recoverable gas in MCF per acre-foot.

Example: 10,500-foot hole with 20 percent porosity:

$$2 \times 20 \times 50 = 2,000 \text{ MCFG/acre-foot}$$

A similar rule for oil would be 10 x porosity for barrels per acre-foot, based on a 25 percent recovery factor.

§ 7.05 Economics

There are nearly as many methods for determining the profitability of an exploration project as there are companies in the business. Some are quite simple and some extremely complex, apparently depending on access to a computer. The goal, of course, is to determine whether, if successful in finding hydrocarbons, the project will make money. Or, to turn the idea around, one might want to determine the minimum figure for reserves that must be discovered with the given investment for the project to be worth pursuing.

The following is a list of parameters which should be included in any profitability analysis:

1. Acreage costs
2. Dry hole costs
3. Completion costs
4. Development well costs

Usually included in the acreage costs are the costs of geology, geophysics, finder's fees, rentals and administrative overhead. All of these factors, added together, represent the total investment if the project is a success. Against these costs the potential return is compared. Beginning with the total expected reserves, each of the following parameters must be subtracted:

1. Royalty
2. Overriding royalty
3. Taxes
4. Operating expenses and overhead
5. Workover costs

At this point the number, in dollars, represents the net return over the life of the project. It is the undiscounted net income. The next step is to apply a discount factor to take into consideration the delay in the return on the investment. The discount rates used will vary according to the outlook of the company but the range is usually from five to 12 percent.

In general, a drilling deal which carries a low geological risk usually has a low rate of return also. This is the so-called "close-in deal" so popular in recent years. A minimum rate of return on such an investment should be two to one. For a rank wildcat, with a high geological risk, the estimated rate of return should be at least 10 to one.

To summarize, a prospect should be based on sound geological concepts with careful subsurface geology and, if possible, geophysics in support. If similar production has already been established in the region, or if there are shows of oil and gas in nearby dry holes, the prospect will be enhanced.

Appendix One

Sources of Geological Information

Federal Agencies

U.S. Geological Survey
  For maps of areas west of the Mississippi:
    Distribution Section
    U.S. Geological Survey
    Federal Center
    Box 25286
    Denver, Colorado 80225
  For maps of areas east of the Mississippi and for all U.S.G.S. books:
    Distribution Section
    U.S. Geological Survey
    1200 South Eads Street
    Arlington, Virginia 22202

Some U.S.G.S. offices have "over-the-counter" sales of both books and maps. You can consult catalogs of U.S.G.S. publications at public libraries as well as university libraries. You must order by name of publication as well as stock number and pay in advance. Maps can be expected in two to three weeks. Reports may take months.

For topographic maps, if you are located in the area of interest you will find state geological survey offices as well as other commercial map services offering "topo sheets" for sale. These are sometimes slightly higher in price ($1.50 vs. $1.25), but are immediately available.

Aerial Photos

Two government agencies offer aerial photos for sale in two forms — the photo index sheet at a scale, usually of 1 inch = 1 mile, and contact prints at a scale of 1:17,000 to 1:30,000. The U.S. Department of Agriculture and the U.S. Geological Survey each have photos available. You can examine photos at the County Office of the Soil Conservation Service before ordering. Ask for "Status of Aerial Photography Coverage" and request order forms. Prepayment is required.

ERTS Pictures

For reconnaissance work, Earth Resources Technology Satellite "photos" can be useful. They are not really photographs, but are images which look very much like photos. Two satellites are in polar orbit around the earth at an altitude of about 570 miles. The satellites complement each other and repeat each image every nine days. Each frame of the imagery (now called Landsat Imagery) covers an area 115 x 115 statute miles and the images are available in several scales and filters. For index maps and other information contact:

Earth Resources Observation System (EROS) Data Center
Sioux Falls, South Dakota 57198

Soil Surveys

Current surveys are published at a scale of 1:20,000 on a photo base. They are sometimes available free at the Soil Conservation Service office in each county. Write for a list of available surveys to the U.S. Department of Agriculture.

State Agencies

Most states have a geological survey, although it may have a different name, such as Nevada Bureau of Mines and Geology. They are usually located in the state capital and they will have numerous maps and reports for sale. You can write or call saying, "Send me what you have on Green County (or other local area)," and they are usually very good about responding. Other state agencies which can be good sources of information include the regulatory agencies and tax departments.

147

National Geological Organizations

The American Association of Petroleum Geologists (AAPG) is the largest single organization of geologists and is international in scope, despite its name. *The AAPG Bulletin*, published monthly, contains articles of wide range which are usually highly technical but can be of value to the non-geologist. The AAPG also publishes special volumes of papers and some maps. Three indexes have been compiled, for the years 1946-1955, 1956-1965 and 1966-1970. Annual indexes are in each December issue. For information contact:

American Association of Petroleum Geologists
P.O. Box 979
Tulsa, Oklahoma 74101

The Geological Society of American (GSA) is a major national organization which does not limit itself to any specialty. *The Bulletin* contains highly technical materials and many special publications, such as the *GSA Memoirs*, are available. For more information, contact:

Geological Society of America
P.O. Box 1719
Boulder, Colorado 80302

The American Institute of Professional Geologists (AIPG) is an organization devoted to professional problems rather than scientific problems. These include relationships of geologists to federal and state legislation, employer-employee relations, professional practice, and ethical standards. The membership directory lists the specialties of each geologist and geophysicist. For more information, contact:

American Institute of Professional Geologists
P.O. Box 957
Golden, Colorado 80401

Local Geological Societies

The major cities in oil and mining provinces usually have geological societies or associations and they frequently publish good geological material, including field trip guidebooks. They can direct you to reputable local geological consultants and they sometimes maintain libraries of electric logs and scout tickets which can be examined for a small fee. A compilation in *Geotimes* for July/August, 1977 lists some 340 "Societies in Earth Science" spread throughout the world.

Other

In addition to the sources listed above, there are trade journals, commercial map companies, petroleum directors, and electric log services available in most oil communities.

There are also many libraries having superior collections of earth science material. Most of these are located at state universities.

References:

Directory of Geological Material in North America by J.V. Howell and A.I. Levorsen; second edition revised and enlarged with the assistance of R.H. Doll and Jane Weaver wilds; American Geological Institute, Washington, D.C., 1957.

Geologic Reference Sources by D.C. Ward, M.W. Wheeler, and M.W. Pangborn, Jr.; The Scarecrow Press, Inc., 1972.

Availability of Earth Resources Data; U.S. Geological Survey, Inf. 74-30.

Appendix Two

General References

Dictionary of Geological Terms, Dolphin Books, paperback.

Oil From Prospect to Pipeline by Robert R. Wheeler and Maurine Whited. Gulf Publishing Co. Third Edition, 1975.

Subsurface Mapping by M.S. Bishop. John Wiley & Sons, 1960.

Mapping by David Greenhood. Phoenix Science Series, 1964, paperback.

Petroleum Geology of the United States by Kenneth K. Landes. Wiley Interscience, 1970.

Geology Illustrated by John S. Shelton. W.H. Freeman & Co., 1966.

D & D Standard Oil Abbreviator compiled by Desk & Derrick Clubs of North America. The Petroleum Publishing Co. Second Edition.

Appendix Three

Calculation for Recoverable Oil and Recoverable Gas

Calculation For Recoverable Oil

The formula for calculating primary recoverable oil requires the parameters listed below. The result is in barrels per acre-foot.

1. Original volume (7,758)
2. Porosity (percent)
3. Oil in place (percent)
4. Recovery factor (percent)

Example

7,758 x .15 x .55 x .20 = 128 B/acre-foot

To complete the calculation, multiply 128 by the number of acre-feet which are assumed to be productive. If the pay is 20 feet thick and the well will drain 160 acres, then:

128 x 20 x 160 = 409,600 Barrels of oil

This is the expected ultimate primary recovery of one well.

Discussion

Of the six parameters needed to calculate the ultimate recovery, the first is a constant — there are 7,758 barrels in an acre-foot. Porosity can be taken from core analysis or log analysis. The amount of oil in place is usually based on electric log calculations of the amount of water in place. If there is 45 percent water (Sw) then the oil in place is 55 percent. The recovery factor is usually based on experience and empirical data. Twenty percent is quite common. Special reservoir characteristics, such as water-drive rather than a solution gas-drive, will alter the recovery factor. Reservoir engineers will be able to place a realistic recovery on most fields which occur in areas of already-established production.

Calculation for Recoverable Gas

This calculation is much more complicated than the one for oil because many additional factors must be taken into consideration. The basic parameters needed are:

1. Original volume (43,560), (a constant)
2. Porosity (percent)
3. Gas in place (percent)
4. Pressure factor
5. Temperature factor
6. Recovery factor

Each of the last three factors has to be calculated separately from bottom-hole information plus knowledge of compressibility factors, abandonment pressures, etc. Each new factor refines the calculation.

Example

43,560 x .20 x .55 x 50.68 x .928 x .80 + 180,283 MCF/acre-foot

As with the oil calculation, one must multiply the answer by the net feet of pay and the drainage area to get the ultimate recovery of one well. If the well in this example has 40 feet of pay and will drain 640 acres, then:

180,283 x 40 x 640 = 4,615,244,800 or 4.6 billion cubic feet of gas

Rules of Thumb

Because of the complexity of the gas calculation, many people use some handy rules of thumb which can usually produce a fair estimate. Here are two:

1. Depth in feet x 100 / 1,000 = MCF/acre-foot

2. Porosity x 30 x Depth in miles = MCF/acre-foot
   Example: .18 x 30 x 5,500/5,280 = 567 MCF/acre-foot

Appendix Four

Common Symbols

COMMON MAP SYMBOLS

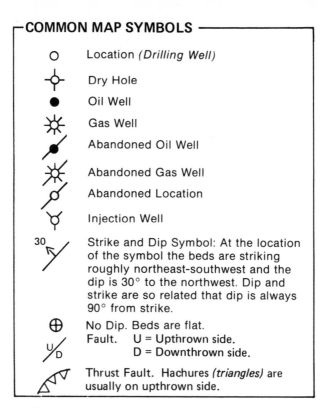

○	Location *(Drilling Well)*
	Dry Hole
●	Oil Well
☼	Gas Well
	Abandoned Oil Well
	Abandoned Gas Well
	Abandoned Location
	Injection Well

30 Strike and Dip Symbol: At the location of the symbol the beds are striking roughly northeast-southwest and the dip is 30° to the northwest. Dip and strike are so related that dip is always 90° from strike.

⊕ No Dip. Beds are flat.

U/D Fault. U = Upthrown side.
 D = Downthrown side.

Thrust Fault. Hachures *(triangles)* are usually on upthrown side.

COMMON CROSS SECTION SYMBOLS

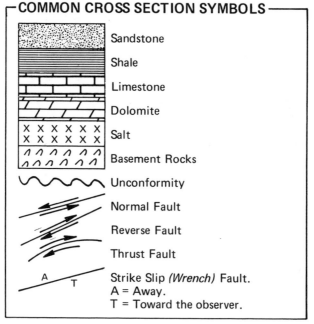

	Sandstone
	Shale
	Limestone
	Dolomite
	Salt
	Basement Rocks
	Unconformity
	Normal Fault
	Reverse Fault
	Thrust Fault
	Strike Slip *(Wrench)* Fault. A = Away. T = Toward the observer.

Appendix Five

Geologic Time Scale

ERA	PERIOD	EPOCH	AGE*
CENOZOIC	Quaternary	Recent (Holocene)	1
		Pleistocene	2
	Tertiary	Pliocene	12
		Miocene	26
		Oligocene	37
		Eocene	53
		Paleocene	65
MESOZOIC	Cretaceous		136
	Jurassic		190
	Traissic		225
PALEOZOIC	Permian		280
	Pennsylvanian		320
	Mississippian		345
	Devonian		395
	Silurian		430
	Ordovician		500
	Cambrian		570
PRECAMBRIAN			4,500

* AGE (in millions of years)

Appendix Six

Units of Measurement and Metric Conversion

1 inch	=	2.54 centimeters
1 foot	=	0.305 meters
1 yard	=	0.91 meters
1 mile	=	1.61 kilometers

— OR —

1 centimeter	=	0.39 inches
1 meter	=	3.28 feet
1 kilometer	=	3,280 feet or 0.62 miles
1 acre	=	0.40 hectares
1 square mile	=	2.6 square kilometers
ɛgrees Celsius	=	Degrees Fahrenheit - 32 / 1.8
es Fahrenheit	=	Degrees Celsius x 1.8 + 32

INDEX

A

B

C

165

E

F

166

M

R

S

U

V

W